ANATOMICAL TERMS

AND THEIR DERIVATION

ANATOMICAL TERMS
AND THEIR DERIVATION

F. PETER LISOWSKI
University of Tasmania, Australia

CHARLES E. OXNARD
University of Western Australia, Australia

World Scientific

NEW JERSEY • LONDON • SINGAPORE • BEIJING • SHANGHAI • HONG KONG • TAIPEI • CHENNAI

Published by

World Scientific Publishing Co. Pte. Ltd.

5 Toh Tuck Link, Singapore 596224

USA office: 27 Warren Street, Suite 401-402, Hackensack, NJ 07601

UK office: 57 Shelton Street, Covent Garden, London WC2H 9HE

British Library Cataloguing-in-Publication Data
A catalogue record for this book is available from the British Library.

ANATOMICAL TERMS AND THEIR DERIVATION

ISBN   981-270-387-X (pbk)

Typeset by Stallion Press
Email: enquiries@stallionpress.com

Printed by FuIsland Offset Printing (S) Pte Ltd, Singapore

CONTENTS

ANATOMICAL TERMS AND THEIR DERIVATION

Anatomy is gradually disappearing from the medical curriculum. This is partly because the curriculum is overcrowded and something has to go. It is also because medical education assumes that anatomy's main value is for surgery and will be learnt later by beginning surgeons. Yet anatomy for surgery is, paradoxically, the least of the reasons for doctors to know anatomy. There is, thus, a minor "surgery" of general practice that requires anatomy in situations where it cannot be "looked up." The knife for a sebaceous cyst in the neck easily leads to paralysis of the trapezius if anatomical knowledge has gone (or never existed). The needle at the elbow or buttock readily causes problems if the simple anatomy is absent (or has disappeared). Even the finger (touching — palpation) or the eye (looking — observation) mislead if what is under the skin has been forgotten (or was never known).

Thus, the use of anatomy is widely misunderstood; it is certainly far wider than just surgery. Anatomy helps understand signs and symptoms. Do we know how the local anatomy of special skin regions relates to oedema: swelling of the back of hand and foot in the upright patient, swelling around the eyes in the lying patient, swelling of the genitalia (especially alarming to the patient in a cardiac bed) in the sitting patient? Anatomy is important in clinical tests. Do we understand: the opposite side test, pressing in left iliac fossa giving pain in right iliac fossa in appendicitis; the obturator test, flexing and medially rotating the right thigh causes pain from a pelvic appendix lying on the obturator muscle? Anatomy is important in everything we do in medicine. Do we know that accuracy in observation and description stem first from observation and description in anatomy? Do we realize that anatomical terms are the vocabulary of medicine, not just anatomy: how doctors communicate with one another, and with the patient?

Learning anatomy is also widely misunderstood; it is certainly far wider than just memorization of facts. Of course, memorization is an important aid to learning in all disciplines. Paradoxically, this is one of the reasons for eliminating anatomy; it seems that it does not need to be learnt; it can be "looked up." However, the information in texts, lists, pictures, notes, mnemonics and memory (rote-learning) though all useful initially, disappear in short order. Anatomy is only truly known through understanding the underlying science; development: how it came to be during a life time; evolution: how it came to be in myriads of life times; function: how it works; experience: how it is used; application: when it goes wrong.

Anatomy can be learnt from text-books, atlases, models, prosections and computers. This is second hand learning.

It can also be learnt from many bodies, of the living as well as the dead, of teachers, fellow students, health care colleagues, patients, and of one's own body as well as the patient's. This is powerful first hand learning: listening, speaking, discussing, using, and above all, questioning. Questioning is the final basis of learning. All this communication requires an interface between the language of medicine and the language of life.

All this leads to a new problem for today's students. In prior times, students were often relatives of doctors or other professionals already possessing a medical flavour; they usually had English as their first language; they had biological and classical backgrounds; they already "knew" what the words meant. In today's world, in contrast, many students are from language groups (Asia, Africa, Eastern Europe, the Middle East) far removed from English; many lack the classics and biology that utilize the words (in Australia, biological sciences are not even required at entry); many are baffled by the language of medicine. Such students ask:

"Why do things have such complex names?"
"If only we knew the meaning it would be easier to understand!"

The teacher of the old days who could have enlightened them is almost extinct. The following text provides some of the information that those teachers might have given. It has been tested on medical students in the Universities of Tasmania and Western Australia. The student response has been most gratifying. The small pocket word book is truly a *vade mecum* that goes with the student into all parts of medicine.

Anatomy began as descriptive science in the days when Latin was the universal scientific language, and early anatomists described the structures they saw in that language, comparing them to common and familiar objects, or borrowing terms from the Greek and Arabic masters before them.

In anatomic terminology, common Latin or Greek words are used as such for any part of the body for which the ancients had a name. For many other structures, scientific names have been invented (1) by using certain classical words which appear to be descriptive of the part concerned, or (2) commonly, by combining Greek or Latin roots to form a new compound term. Memorization of such terms without understanding their meaning can lead to mental indigestion. Here, the roots are also presented from which many of these descriptive terms and compounds are derived, as an aid to comprehension. For practical convenience, the book is organised into abbreviations, prefixes and suffixes, general terms common to all body regions, short lists for each major body part, and an alphabetical list covering the entire body.

For a wider vocabulary, the use of a standard biological or medical dictionary is recommended.

Our thanks go to Ms Jill Aschman, Secretary of the Anatomy and Physiology Department who did the original word processing, and to Dr K. K. Phua and Ms Lim Sook Cheng of World Scientific. Most important thanks of all go to generations of medical students in all the medical schools in which we have worked. Their questions about meanings have stimulated our answers and this pocket book.

Hobart and Perth   FPL
                                                                    CEO
                                                                    2006

ABBREVIATIONS

AS. Anglo-Saxon
A. Arabic
dim. diminutive
F. French
Ger. German
G. Greek
L. Latin
NL. "New" Latin
OF. Old French
pl. plural

COMMONLY USED PREFIXES

Prefix	Meaning	Example
a-, an-	G. lacking, without	asexual, without sex
ab-	L. away from	abduction, lead away
ad-	L. to, toward	adduction, lead towards
af-	L. to, toward	afferent, bring towards
amphi-	G. on both sides, double	amphiarthrosis, two joints
andro-	G. a male	androgen, male producing
ante-	L. forward, before	antebrachium, forearm
anti-	G. against, reverse, opposed to	antihelix, reverse coil
apo-	G. from	apocrine, secreting from
auto-	G. autos, self	autonomic, self governing
basi-	L. pertaining to the base	basiocciput, occipital base
bi-	L. twice, double	bicipital, two heads
brachy-	G. short	brachycephaly, short head
cephalo-	G. head	cephalometry, head measure
circum-	L. around, about	circumduction, lead around
chondro-	G. *chondras*, cartilage	chondrocyte, cartilage cell

Commonly Used Prefixes

co-, com-, con-	L. together, with	constrictor, draw together
de-	L. away from, down	deglutition, swallow down
di-	G. double, twice	diaplegia, both sides
di-, dia-	G. through, between	diaphragm, wall between
dis-	L. apart, away	disarticulate, joint apart
e-, ecto-, ex-	G. on outer side	ectoderm, outer layer
extra-	L. *extra*, outside	extradural, outside the dura
em-, en-	G. in, on	embolism, wedge inside
en-, endo-	G. within	endocardium, within heart
epi-	G. upon, over	epicardium, over heart
eu-	G. the better of alternates	eugenics, better generation
exo-	G. outside	exophthalmos, bulging eye
extra-	L. outside	extracellular, outside the cell
gastro-	G. *gaster*, stomach, belly	gastrocolic reflex, stomach-colon reflex
helico-	G. twisted	helicotrema, twisted hole
hemi-	G. half	hemisphere, half a sphere
holo-	G. whole	holocrine, whole secretion
homo-	G. one and the same	homologous
hyper-	G. over, above, excessive	hypertrophy, over growth
hypo-	G. under, deficient, below	hypothalamus, below the thalamus
im-, in-	L. not	immature, not mature
infra-	L. below	infraorbital, below orbit

3

inter-	L. between	intervertebral, between the vertebrae
intra-	L. inside, within	intramuscular, within muscle
intro-	L. within, into	introspection, look inside
iso-	G. equal	isometric, equal measure
macro-	G. big, large	macrocyte, large cell
mega-	G. great	megalocyte, large cell
meta-	G. change, after	metacarpal, after the carpus (wrist)
mes-, meso-	G. middle	mesogastrium
micro-	G. small	microscope, looking small
myo-	G. *mys*, muscle	combining form of muscle
neuro-	G. *neuron*, nerve	combining form of nerve (tendon, see aponeurosis)
omni-	L. *omnis*, all	omnipotent, all powerful
omphalo-	G. combining form (*omphalos*)	signifying the navel
oo-	G. *oon*, egg	oocyte, egg cell
ortho-	G. *orthos*, straight, direct	orthograde, straight walking
osteo-	G. *osteon*, bone	osteocyte, bone cell
para-	G. by the side of	paravertebral, beside the vertebrae
peri-	G. around	peritoneum, stretched around
post-	L. after, behind	postaxial, behind the axis
prae-, pre-	L. before	prenatal, before birth
pro-	G. before, for	pronephros, before kidney

re-	L. again, back	reflect, turn back
retro-	L. backward, behind	retroperitoneal, behind the peritoneum
semi-	L. half	semimembranosus, half membranous
sub-	L. under	subclavius, under clavicle
supra-	L. above	supraspinatous, above spine
sym-, syn-	G. together, with	symphysis, synostosis
trans-	L. across	transpyloric, across pylorus
ultra-	L. in excess, beyond	ultrasonic, beyond (audible) sounds

COMMONLY USED SUFFIXES

Suffix	Meaning	Example
-algia	G. *algos*, pain	neuralgia, nerve pain
-coele	G. *koilia*, hollow	blastocoele, hollow cell ball
-ectomy	G. *ek-tome*, cutting out	appendicectomy
-form	L. *forma*, shape	fusiform, spindle shaped
-graph	G. *graphein*, to write	radiograph
-itis	G. suffix indicating inflammation	conjunctivitis
-logy	G. *logos*, treatise, word, study of	pathology
-oid	G. *eidos*, form, resemblance	mastoid, breast shaped
-opia	G. *opsis*, sight	diplopia, double vision
-pathy	G. *pathos*, suffering	sympathy
-rhoea	G. *rhein*, to flow	diarrhoea
-scopy	G. *skopein*, to view	endoscopy
-tomy	G. *temnein*, to cut	hysterectomy
-uria	G. *ouron*, urine	haematuria, blood in urine

TERMS COMMON TO ALL
ANATOMICAL REGIONS

Abductor.	L. *ab*, away, + *ducere*, to lead.
Aberrant.	L. *ab*, away, + *errare*, to stray.
Accessory.	L. *accessorius*, supplementary.
Adductor.	L. *ad*, to, + *ducere*, to lead.
Adhesion.	L. *adhaereo*, to stick together.
Adipose.	L. *adeps*, fat.
Aditus.	L. *aditus*, opening.
Adnexa.	L. *ad*, to, + *nexus*, bound.
Adventia.	L. *ad*, to, + *venire*, to come.
Afferent.	L. *ad*, to, + *ferre*, to carry.
Agonist.	G. *agonistes*, rival.
Analogy.	G. *ana*, according to, + *logos*, treatise.
Anastomosis.	G. *anastomoein*, to bring to a mouth, cause to communicate.
Anatomy.	G. *ana*, apart, + *tennein*, to cut.
Anlage.	Ger. *an*, on, + *legen*, to lay (a laying on — *primordium*, precursor).
Annulus.	L. *anulus* (*annulus*), a ring.
Anomaly.	G. *an*, without, + *nomos*, law.
Antagonist.	G. *anti*, against, *agonistes*, rival.
Anteflexion.	L. *ante*, before, + *flexere*, to bend.
Anterior.	L. *ante*, before.
Anteversion.	L. *ante*, before, + *versio*, turning.
Apertura.	L. *apertura*, opening.
Apocrine.	G. *apo*, from, + *krinein*, to separate.

Aqueous.	L. *aqua*, water.
Arrector.	L. *arrigere*, to raise, thus arrectores pilorum, small muscles in skin that erect the hairs.
Artefact.	L. *arte*, by art, + *factus*, made.
Artery.	G. *aer*, air, + *terein*, to keep (L. *arteria*, windpipe); ancient belief that blood vessels contained air.
Articulation.	L. *artus*, joint, *articulatus*, little joint, pl. *articulationes*, joints.
Aspera.	L. *asper*, rough.
Asthenic.	G. *a*, without, + *sthenos*, strength.
Atavistic.	L. *atavus*, grandfather.
Ataxia.	G. *a*, without, + *taxis*, order.
Atelectasis.	G. *ateles*, incomplete, + *ectasis*, expansion.
Atresia.	G. *a*, without, + *tresis*, hole.
Atrophy.	G. *a*, without, + *trophe*, nourishment.
Autonomic.	G. *autos*, self, + *nomos*, law.
Axial.	L. *axis*, axle of a wheel, the line about which any body turns.
Axis.	L. *axis*, axle of a wheel.
Axon.	G. *axon*, axis.
Basal.	L. *basis*, footing, base.
Bifid.	L. *bis*, two, + *findere*, to cleave.
Bifurcate.	L. *bis*, two, + *furca*, fork.
Bilateral.	L. *bi*, two, + *latus*, side.
Biventer.	L. *bis*, two, + *venter*, belly.
Brevis.	L. *brevis*, short.
Bulbus.	L. *bulbus*, bulb, swollen root.
Bulla.	L. *bulla*, a bubble; hence spherical in shape.
Bursa.	L. *bursa*, a purse; hence purse-shaped object.

Calcar.	L. *calcar*, spur.
Calcification.	L. *calx*, lime, + *facere*, to make.
Callous.	L. as above, newly formed bone at fracture site.
Canaliculus.	L. *canalis*, watepipe.
Cancellous.	L. *cancelli*, latticework.
Capillary.	L. *capillaris*, pertaining to the hair.
Capsule.	L. dim. of *capsa*, box.
Cardinal.	L. *cardinalis*, pertaining to a door hinge, on which something important or fundamental hinges.
Cartilage.	L. *cartilago*, gristle, cartilage.
Caruncle.	L. dim. of *caro*, flesh, any fleshy eminence.
Cauda, caudal	L. *cauda*, tail.
Cava.	L. *cavus, -a, -um*, hollow or cave.
Cavernosus.	L. *caverna*, a hollow or cave.
Cavum.	L. *cavum*, a hollow or cave.
Chorda.	G. *chorde*, string of gut, cord.
Chorion.	G. *chorion*, skin.
Choroid.	G. *chorion*, skin, + *eidos*, shape, likeness.
Circumflex.	L. *circum*, around, + *flexere*, to bend.
Collagen.	G. *kolla*, glue, + L. *gen*, begetter of.
Collateral.	L. *con*, together, + *latus*, side.
Colliculus.	L. *colliculus*, little hill.
Colloid.	G. *kolla*, glue, + *eidos*, likeness, shape.
Comes.	L. *comes*, companion.
Comitans (pl. comitantes).	L. *comitari*, to accompany.
Commissure.	L. *commissura* (*cum* + *mittere*), connection.
Communicans.	L. *communicans*, communicating.
Convolution.	L. *con*, together, + *volvo*, to roll.
Corpus (pl. corpora).	L. body.

9

Corpuscle.	L. *corpusculum*, little body.
Crista.	L. *crista*, crest.
Cruciate.	L. *crux*, cross.
Crus (pl. **crura**).	L. *crus*, leg.
Cutaneous.	L. *cutis*, skin.
Cuticle.	L. *cutis*, skin.
Cutis.	L. *cutis*, skin.
Cyst (adj. **cystic**).	G. *kystis*, bag, bladder, pouch.
Decussation.	L. *decussatio*, intersection of two lines, as in Roman X.
Deferens.	L. *de*, away, + *ferens*, carrying.
Dendrite.	G. *dendros*, tree.
Depressor.	L. *de*, down, + *premere*, to press.
Dermal.	G. *derma*, skin.
Dermatome.	G. *derma*, skin, + *temnein*, to cut.
Dermis.	G. *derma*, skin.
Detritus.	L. *deterere*, to rub away.
Detrusor.	L. *detrudere*, to push down.
Distal.	L. *distare*, to stand apart.
Diverticulum.	L. *divertere*, to turn aside.
Dizygotic.	G. *dis*, twice, + *zygoo*, join together.
Dorsal.	L. *dorsum*, back.
Duct.	L. *ducere*, to lead or draw.
Ductule.	L. *ducere*, to lead, dim.
Ectoderm.	G. *ektos*, outside, + *derma*, skin.
Effector.	L. *efficere*, to bring to pass.
Efferent.	L. *ex*, out, + *ferre*, to bear.
Eminence.	L. *e*, out, + *minere*, to jut.
Emissary.	L. *e*, out, + *mittere*, to send.
Endochondral.	G. *endon*, within, + *chondros*, cartilage.
Endoderm.	G. *endon*, within, + *derma*, skin.

Endomysium.	G. *endon*, within, + *mys*, muscle.
Endoneurium.	G. *endon*, within, + *neuron*, nerve.
Endosteum.	G. *endon*, within, + *osteon*, bone.
Endoskeleton.	G. *endon*, within, + *skeletos*, dried up.
Endothelium.	G. *endon*, within, + *thele*, nipple.
Epidermis.	G. *epi*, on, + *derma*, skin.
Epimysium.	G. *epi*, upon, + *mys*, muscle.
Epineurium.	G. *epi*, on, + *neuron*, nerve.
Epithelium.	G. *epi*, on, + *thele*, nipple.
Eversion.	L. *e*, out, + *vertere*, to turn.
Exocrine.	G. *ex*, out, + *krinein*, to separate.
Extension.	L. *extendo*, extend.
Extensor.	L. *ex*, out, + *tendere*, to stretch.
Extrinsic.	L. *extrinsecus*, on the outside.
Exudate.	L. *ex*, out, + *sudare*, to sweat.
Fascia (pl. -iae).	L. *fascia*, band.
Fasciculus.	L. *fascis*, bundle.
Fenestra.	L. *fenestra*, window.
Fibre, Fiber.	L. *fibra*, fibre, string, thread.
Fibril.	NL. *fibrilla*, a little thread.
Fibroblast.	L. *fibra*, fibre, + *blastos*, bud.
Fibrocartilage.	L. *fibra*, fibre, + *cartilago*, gristle.
Filament.	L. *filamentum*, thin fibre.
Fissure.	L. *fissura* (*findo*), a cleft.
Fistula.	L. *fistula*, pipe.
Flaccid.	L. *flaccidus*, weak.
Flexor.	L. *flexus*, bent.
Follicle.	L. *folliculus*, a small bag.
Foramen.	L. *foramen*, hole.
Form.	L. *forma*, shape.
Fossa.	L. *fossa*, ditch, channel, something dug.
Fundiform.	L. *funda*, sling, + *forma*, shape, likeness.

Fundus.	L. *fundus*, bottom.
Funicular.	L. *funis*, rope.
Fusiform.	L. *fusus*, spindle, + *forma*, shape, likeness.
Germinal.	L. *germen*, bud, germ.
Germinative.	L. *germen*, bud, germ.
Gestation.	L. *gestare*, to bear.
Gland.	L. *glandula*, dim. *glans*, acorn, pellet.
Glia.	G. *gloia*, glue.
Habitus.	L. *habitus*, condition of the body.
Haemal, hemal.	G. *haima*, blood.
Hemiplegia.	G. *hemi*, half, + *plege*, stroke.
Hemisphere.	G. *hemi*, half, + *sphaira*, ball.
Hemoglobin, Haemoglobin.	G. *haima*, blood, + L. *globus*, sphere.
Hemopoietic, Haemopoietic.	G. *haima*, blood, + *poietikos*, creative.
Hilum.	L. *hilum*, a small thing.
Holocrine.	G. *holos*, whole, + *krinein*, to separate.
Homology.	G. *homos*, same, + *logos*, treatise.
Hormone.	G. *hormaein*, to excite.
Humor.	L. *humor*, moisture, fluid.
Hypaxial.	G. *hypo*, under, + L. *axis*, centre line, axis.
Implantation.	L. *in*, in, + *plantere*, to plant.
Impression.	L. *in*, in, + *premere*, to press.
Incisura.	L. *incidere*, to cut into.
Inductor.	L. *inducere*, to lead on, excite.
Innervation.	L. *in*, in, + *nervus*, nerve.
Insertion.	L. *in*, in, + *serere*, to plant.
Integument.	L. *in*, over, + *tegere*, to cover.

Intercalated.	L. *inter*, between, + *calare*, to call, inserted or placed between.
Interdigitating.	L. *inter*, between, + *digitus*, digit, interlocked by finger-like processes.
Interstitial.	L. *inter*, between, + *sistere*, to set, thus placed in spaces or gaps.
Intima.	L. *intima*, innermost.
Intrinsic.	L. *intrinsecus*, inward.
Invagination.	L. *in*, in, + *vagina*, sheath.
Inversion.	L. *in*, in, + *vertere*, to turn.
Juxtaposition.	L. *juxta*, near, + *positio*, place.
Karyocyte.	G. *karyon*, nucleus, nut, + *kytos*, cell.
Karyolysis.	G. *karyon*, nucleus, nut, + *lysis*, loosening.
Kinetic.	G. *kinesis*, movement.
Lacerate.	L. *lacerare*, to tear.
Lacuna.	L. *lacuna*, pond.
Lamella.	L. dim. of *lamina*, leaf.
Lamina (pl. -ae).	L. *lamina*, thin plate.
Lateral.	L. *lateralis* (*latus*), pertaining to a side.
Leucocyte, Leukocyte.	G. *leukos*, white, + *kytos*, cell.
Ligamentum.	L. *ligamentum*, a bandage.
Lipid.	G. *lipos*, fat, + *eidos*, resemblance.
Lobus.	G. *lobos*, lobe.
Locus.	L. *locus*, place.
Lymph.	L. *lympha*, spring water.
Lymphocyte.	L. *lympha*, spring water, + G. *kytos*, cell.
Macrophage.	G. *makros*, large, + *phagein*, to eat.
Macroscopic.	G. *makros*, large, + *skopeo*, I see.

Marginal.	L. *marginalis* (*margo*), bordering.
Matrix.	L. *matrix* (*mater*), womb, groundwork, mold.
Meatus (pl. -us).	L. *meatus*, passage.
Medial.	L. *medialis* (*medius*), pertaining to the middle.
Median.	L. *medianus*, in the middle.
Megaloblast.	G. *megas*, big, + *blastos*, bud.
Membrane.	L. *membrana*, skin.
Merocrine.	G. *meros*, portion, + *krinein*, to separate.
Mesenchyme.	G. *mesos*, middle, + *en*, in, + *chymos*, juice.
Mesoderm.	G. *mesos*, middle, + *derma*, skin.
Mesothelium.	G. *mesos*, middle, + *thele*, nipple, hence middle lining layer.
Metabolic.	G. *metabole*, change.
Metaphase.	G. *meta*, after, + *phasis*, appearance.
Metaplasia.	G. *meta*, after, + *plasma*, formed, molded.
Microscope.	G. *mikros*, small, + *skopeo*, I look.
Microsome.	G. *mikros*, small, + *soma*, body.
Mitochondria.	G. *mitos*, thread, + *chondrion*, grain.
Mitosis.	G. *mitos*, thread.
Monocyte.	G. *monos*, single, + *kytos*, cell.
Morphology.	G. *morphos*, form, + *logos*, treatise.
Mucus.	L. *mucus*, G. *muxa*, snivel, slippery secretion.
Muscle.	L. *musculus*, little mouse.
Myology.	G. *mys*, muscle, + *logos*, treatise.
Myotome.	G. *mys*, *myos*, muscle, + *tome*, a cutting.
Necrosis.	G. *nekrosis*, a killing.
Necropsy.	G. *nekros*, corpse, + *opsis*, sight.
Nerve.	L. *nervus*, G. *neuron*, cordlike structure, nerve, tendon.
Neural.	G. *neuron*, nerve.
Neuralgia.	G. *neuron*, nerve, + *algos*, pain.

Neuraxon.	G. *neuron*, nerve, + *axon*, axis.
Neurilemma.	G. *neuron*, nerve, + *lemma*, husk, sheath.
Neuroblast.	G. *neuron*, nerve, + *blastos*, bud.
Neuroectomy.	G. *neuron*, nerve, + *ektome*, excision.
Neuroglia.	G. *neuron*, nerve, + *gloia*, glue.
Neurology.	G. *neuron*, nerve, + *logos*, treatise.
Neuron.	G. *neuron*, cordlike structure, sinew, tendon; equivalent of L. *nervus*, hence, nerve.
Neuropil.	G. *neuron*, nerve, + *pilos*, felt.
Neuropore.	G. *neuron*, nerve, + *poros*, hole.
Node.	L. *nodus*, knot.
Norma.	L. *norma*, rule, square used by carpenters, hence standard viewpoint.
Normoblast.	L. *norma*, rule, + *blastos*, bud.
Nucleolus.	L. *nucleus*, nut. New derivation "small nucleus."
Nucleus.	L. *nucleus*, nut.
Occlusion.	L. *ob*, before, + *claudo*, I close.
Oedema, Edema.	G. *oidema*, swelling.
Oestrus, Estrus.	G. *oistros*, stinging insect, stung into activity at time of heat.
Oligodendroglia.	G. *oligos*, few, + *dendron*, tree, + *gloia*, glue.
Opposition.	L. *ob*, in the way of, + *positus*, placed.
Organ.	L. *organum*, implement.
Orifice.	L. *orificium*, opening.
Os.	L. *os*, bone.
Osseous.	L. *os* (pl. *ossa*), bone.
Ossicle.	L. *ossiculum*, small bone.
Ossification.	L. *os*, bone, + *facere*, to make.
Osteone.	G. *osteon*, bone.
Osteocyte.	G. *osteon*, bone, + *kytos*, cell.
Osteology.	G. *osteon*, bone, + *logos*, treatise.

Osteolysis.	G. *osteon*, bone, + *lysis*, melting.
Ostium.	L. *ostium*, door.
Paradox.	G. *para*, alongside of, + *doxa*, belief.
Paraesthesia,	G. *para*, alongside of, + *aisthesis*, sensation.
Paresthesis.	
Paraganglion.	G. *para*, along, beside, + *ganglion*, knot.
Paralysis.	G. *para*, alongside of, + *lyein*, to loosen.
Paramedian.	G. *para*, alongside of, + *mesos*, middle.
Paraplegia.	G. *para*, alongside of, + *plesso*, I strike.
Parasympathetic.	G. *para*, alongside of, + *sympathetikos*, sympathetic.
Paraxial.	G. *para*, alongside of, + L. *axis*, axle.
Paresis.	G. *paresis*, relaxation.
Pathology.	G. *pathos*, disease, + *logos*, treatise.
Perichondrium.	G. *peri*, around, + *chondros*, cartilage.
Perimysium.	G. *peri*, around, + *mys*, muscle.
Periosteum.	G. *peri*, around, + *osteon*, bone.
Phagocyte.	G. *phagein*, to eat, + *kytos*, cell.
Phenotype.	G. *phainein*, to display, + *typos*, type.
Physic.	G. *physikos*, natural.
Physis.	G. *phyein*, to generate, hence an outgrowth.
Piriform.	L. *pirum*, pear, + *forma*, shape, likeness.
Placode.	G. *plax*, anything flat.
Plantigrade.	L. *planta*, side of foot, + *gradior*, to walk.
Plasma.	G. *plasma*, something formed.
Platelet.	OF. *plate*, plate.
Pleomorphic.	G. *pleon*, more (or many) + G. *morphos*, forms.
Plexus (pl. -us).	L. *plexus*, plaiting, braid.
Poikilocyte.	G. *poikilos*, diversified, + *kytos*, cell.
Polymorphonuclear.	G. *polys*, many, + *morphe*, form, + L. *nucleus*, nut; hence mixed term.

Pore.	L. *porus*, passage.
Porta.	L. *porta* (pl. *-ae*), gate.
Portal.	L. *porta* (pl. *-ae*), gate.
Portio.	L. *portio*, part.
Porus.	G. *poros*, passage.
Primordial.	L. *primordium*, beginning.
Processus.	L. *processus*, going forwards.
Prochordal.	G. *pro*, in front of, + *chorde*, cord.
Profundus.	L. *profundus*, deep.
Progesterone.	G. *pro*, before, + *gerere*, to bear.
Promontory.	G. *promontorium*, mountain ridge.
Proprioceptor.	L. *proprius*, special, + *capere*, to take.
Proptosis.	G. *pro*, before, + *ptosis*, falling.
Prosector.	G. *pro*, before, + L. *secare*, to cut.
Protoplasm.	G. *protos*, first, + *plasma*, form.
Protuberance.	L. *protubero*, I swell.
Proximal.	L. *proximus*, next.
Pyknic.	G. *pyknos*, compact.
Pyknosis.	G. *pyknos*, compact.
Pyramid.	G. *pyramis*, pyramid.
Pyriform.	L. *pirum*, pear, + *forma*, shape.
Quadrangular.	L. *quattuor* four + L. *angulus*, four angles.
Quadratus.	L. *quadratus*, squared.
Quadriceps.	L. *quattuor*, four, + *caput*, head.
Quadrigeminus.	L. *quadrigeminus*, four-fold, four.
Radicle.	L. *radix*, root.
Radix.	L. *radix*, root.
Ramify.	L. *ramus*, branch, + *facere*, to make.
Ramus.	L. *ramus*, branch.
Raphe.	G. *raphe*, seam.
Receptor.	L. *recipere*, to take back, receive.

17

Recess.	L. *recessus*, retreat.
Reflect.	L. *reflecto*, to turn back.
Reticulocyte.	L. *reticulum*, little net, + *kytos*, cell.
Reticulum.	L. *reticulum*, little net.
Retraction.	L. *re*, back, + *trahere*, to draw.
Retractor.	L. *retrahere*, to draw back.
Rostrum (pl. **-a**).	L. *rostrum*, beak.
Rotator.	L. *rotare*, to whirl about.
Sac.	L. *saccus*, sack.
Sacculus.	L. *sacculus*, a little bag.
Sagittal.	L. *sagitta*, arrow; shape of saggital suture including the lambdoid suture.
Sarcolemma.	G. *sarx*, flesh, + *lemma*, husk, skin.
Scala.	L. *scala*, staircase.
Sclerotic.	G. *skleros*, hard.
Sebaceous.	L. *sebum*, tallow, grease.
Sebum.	L. *sebum*, tallow.
Sigmoid.	G. *sigma*, Greek letter, + *eidos*, shape, likeness.
Sinister.	L. *sinister*, left side or unlucky.
Sinus (pl. **-us**).	L. *sinus*, curve, cavity, bosom.
Sinusoid.	L. *sinus*, curve, cavity, + *eidos*, shape, likeness.
Soma.	G. *soma*, body.
Somatic.	G. *soma* (pl. *somata*), body.
Somatopleure.	G. *soma*, body, + *pleura*, side.
Somite.	G. *soma*, body, + suffix *-ite*, indicating origin.
Splanchnic.	G. *splanchna*, viscera.
Spongiosum.	G. *spongia*, sponge.
Spongioblast.	G. *spongia*, sponge, + *blastos*, germ, bud.
Squamo-.	L. *squama*, scale.

Stenosis.	G. *stenosis*, narrowing.
Sthenic.	G. *sthenos*, strength.
Stratum (pl. strata).	L. *stratum*, layer.
Stria.	L. *stria*, furrow.
Striatum.	L. *striatus*, grooved, streaked.
Stroma.	G. *stroma*, blanket.
Sulcus.	L. *sulcus*, furrow.
Suture.	L. *sutura*, seam, sewing together.
Sympathetic.	G. *syn*, together, + *pathein*, to suffer.
Synapse.	G. *syn*, together, + *haptein*, to fasten.
Synarthrosis.	G. *syn*, together, + *arthron*, joint.
Synchrondosis.	G. *syn*, together, + *chondros*, cartilage.
Syncytium.	G. *syn*, together, + *cytos*, cell.
Syndesmosis.	G. *syn*, together, + *desmosis*, band.
Syndrome.	G. *syndrome*, occurrence.
Synergy.	G. *syn*, together, + *ergon*, work.
Synostosis.	G. *syn*, together, + *osteon*, bone.
Synovia.	G. *syn*, together, + *ovum*, egg.
Tendon.	L. *tendere*, to stretch.
Tensor.	L. *tendere*, to stretch.
Teres.	L. *tero*, I grind, rub.
Terminalis.	L. *terminare*, to limit.
Theca.	L. *theca*, envelope, sheath.
Thrombocyte.	G. *thrombos*, lump, + *kytos*, cell.
Thrombus.	G. *thrombus*, lump.
Tome.	G. *tennein*, to cut.
Trabecula (pl. -ae).	L. *trabecula*, a little beam.
Tract.	L. *tractus*, wool drawn out for spinning.
Trophoblast.	G. *trophe*, nourishment, + *blastos*, bud.
Tuberosity.	L. *tuber*, round smooth swelling.
Tunica.	L. *tunica*, undergarment.

Unciform.	L. *uncus*, hook, + *forma*, shape.
Uncinate.	L. *uncinatus*, furnished with a hook.
Uncus.	L. *uncus*, hook.
Valve.	L. *valva*, leaf of door.
Valvula.	L. *valvula*, a little fold, valve.
Vas.	L. *vas*, vessel.
Vascular.	L. *vasculum*, small vessel.
Vein.	L. *vena*, vein.
Ventral.	L. *venter*, belly.
Vertex.	L. *vertex*, whirl, whirlpool.
Vestige.	L. *vestigium*, trace, footprint.
Zona.	L. *zona*, belt, girdle.
Zonula.	L. dim. of *zona*, belt, girdle.
Zygote.	G. *zygoein*, to yoke together.

TERMS IMPORTANT IN THE UPPER LIMB

Abductor.   L. *ab*, away, + *ducere*, to lead.
Accessory.   L. *accessorius*, supplementary.
Acromion.   G. *akron*, height, extremity, + *omos*, shoulder.
Adductor.   L. *ad*, to, + *ducere*, to lead.
Anconeus.   G. *agkon*, elbow.
Angle.   L. *angulus*, angle.
Annulus.   L. *anulus* (*annulus*), a ring.
Ansa.   L. *ansa*, handle.
Antebrachium.   L. *ante*, before, + *brachium*, arm.
Anteflexion.   L. *ante*, before, + *flexere*, to bend.
Anteversion.   L. *ante*, before, + *versio*, turning.
Appendicular.   adjectival form of appendix.
Axilla.   L. *axilla*, armpit.

Basilic.   A. *al-basilic*, inner vein.
Biceps.   L. *bis*, two, + *caput*, head.
Bipennate.   L. *bis*, two, + *pinna*, feather.
Biventer.   L. *bis*, two, + *venter*, belly.
Brachial.   L. *brachialis*, belonging to the arm
                         (upper arm).
Brachium (pl. -ia). L. *brachium*, arm, (upper arm).
Brevis.   L. *brevis*, short.
Bursa.   L. *bursa*, a purse; hence purse-shaped object.

Capitulum.   L. dim. *caput*, small head.

Capsule.	L. dim. of *capsa*, box.
Caput (pl. **capita**).	L. *caput*, head.
Carpus.	G. *karpos*, wrist.
Cervical.	L. *cervix*, neck.
Cingulum.	L. *cingulum*, girdle.
Circumflex.	L. *circum*, around, + *flexere*, to bend.
Clavicle.	L. *clavicula*, a little key.
Collateral.	L. *con*, together, + *latus*, side.
Collum.	L. *collum*, neck.
Comes.	L. *comes*, companion.
Comitans	L. *comitari*, to accompany.
(pl. **comitantes**).	
Communicans.	L. *communicans*, communicating.
Condyle.	G. *kondylos*, knuckle.
Coraco-, corono-.	G. *korax* or *corone*, crow; hence crowlike.
Coracoid.	G. *korax*, crow, + *eidos*, form, likeness.
Corium.	G. *chorion*, skin, leather.
Corpus	L. body.
(pl. **corpora**).	
Crista.	L. *crista*, crest.
Cubitus.	L. *cubitus*, elbow.
Dactyl.	G. *daktylos*, finger.
Deltoid.	G. *delta*, letter in Greek alphabet, triangular-shaped.
Dermatome.	G. *derma*, skin, + *temnein*, to cut.
Diaphysis.	G. *dia*, between, + *physis*, growth.
Diarthrosis.	G. *dia*, through, + *arthroun*, to fasten by a joint.
Digit.	L. *digitus*, finger.
Disc.	L. *discus*, disc.
Distal.	L. *distare*, to stand apart.
Dorsal.	L. *dorsum*, back.

Epicondyle.	G. *epi*, on, + *kondylos*, knuckle.
Epiphysis.	G. *epi*, on, + *physis*, growth.
Epithelium.	G. *epi*, on, + *thele*, nipple.
Eversion.	L. *e*, out, + *vertere*, to turn.
Extension.	L. *extendo*, extend.
Extensor.	L. *ex*, out, + *tendere*, to stretch.
Extrinsic.	L. *extrinsecus*, on the outside.
Fascia (pl. -iae).	L. *fascia*, band.
Fasciculus.	L. *fascis*, bundle.
Flaccid.	L. *flaccidus*, weak.
Flexor.	L. *flexus*, bent.
Foramen.	L. *foramen*, hole.
Fossa.	L. *fossa*, ditch, channel, something dug.
Fusiform.	L. *fusus*, spindle, + *forma*, shape, likeness.
Ganglion.	G. *ganglion*, a swelling under the skin.
Ginglymus.	G. *ginglymus*, hinge.
Glenoid.	G. *glene*, socket, + *eidos*, form, likeness.
Hamatum.	L. *hamatus*, hook-shaped.
Hamulus.	L. dim. *hamus*, hook.
Humerus.	L. *humerus*, shoulder.
Hyaline.	G. *hyalos*, glass.
Hypaxial.	G. *hypo*, under, + L. *axis*, centre line, axis.
Incisura.	L. *incidere*, to cut into.
Index.	L. *index*, pointer.
Infraspinous.	L. *infra*, below, + *spina*, spine.
Insertion.	L. *in*, in, + *serere*, to plant.
Integument.	L. *in*, over, + *tegere*, to cover.
Intercalated.	L. *inter*, between, + *calare*, to call, inserted or placed between.

Interdigitating.	L. *inter*, between, + *digitus*, digit, interlocked by finger-like processes.
Intima.	L. *intima*, innermost.
Intrinsic.	L. *intrinsecus*, inward.
Invagination.	L. *in*, in, + *vagina*, sheath.
Inversion.	L. *in*, in, + *vertere*, to turn.
Juxtaposition.	L. *juxta*, near, + *positio*, place.
Kinetic.	G. *kinesis*, movement.
Labrum.	L. *labrum*, rim.
Lateral.	L. *lateralis* (*latus*), pertaining to a side.
Latissimus.	L. superlative, *latus*, broad, therefore very broad.
Levator.	L. *levare*, to raise.
Ligamentum.	L. *ligamentum*, a bandage.
Locus.	L. *locus*, place.
Longus.	L. *longus*, long.
Lumbrical.	L. *lumbricus*, earthworm.
Magnus, -a, -um.	L. *magnus*, large.
Mammary.	L. *mamma*, breast.
Mammillary.	L. dim. *mammillaris* (*mamma*, *-ae*), little breast.
Manubrium.	L. *manubrium*, handle, hilt (as of a sword).
Manus.	L. *manus*, hand.
Marginal.	L. *marginalis* (*margo*), bordering.
Matrix.	L. *matrix* (*mater*), womb, groundwork, mold.
Metacarpus.	G. *meta*, after, + L. *carpus*, wrist.
Metaphysis.	G. *meta*, after, + *physis*, growth.
Minimus.	L. *minimus*, least.

Nerve.	L. *nervus*, G. *neuron*, cordlike structure, nerve, tendon.
Neural.	G. *neuron*, nerve.
Neuralgia.	G. *neuron*, nerve, + *algos*, pain.
Neuron.	G. *neuron*, cordlike structure, sinew, tendon; equivalent of L. *nervus*, hence, nerve.
Node.	L. *nodus*, knot.
Norma.	L. *norma*, rule, square used by carpenters, hence standard viewpoint.
Olecranon.	G. *olene*, elbow, + *kranion*, head.
Opposition.	L. *ob*, in the way of, + *positus*, placed.
Os.	L. *os*, bone.
Osseous.	L. *os* (pl. *ossa*), bone.
Ossicle.	L. *ossiculum*, small bone.
Ossification.	L. *os*, bone, + *facere*, to make.
Osteology.	G. *osteon*, bone, + *logos*, treatise.
Palma.	L. *palma*, the (open) hand.
Paramedian.	G. *para*, alongside of, + *mesos*, middle.
Parasternal.	G. *para*, alongside of, + *sternon*, chest.
Parasympathetic.	G. *para*, alongside of, + *sympathetikos*, sympathetic.
Paraxial.	G. *para*, alongside of, + L. *axis*, axle.
Pectoral.	L. *pectoralis* (*pectus*), belonging to the breast.
Phalanges.	G. *phalanx*, band of soldiers; singular phalanx, originally denoted whole set of bones of a digit not just one bone.
Pisiform.	L. *pisum*, pea, + *forma*, shape.
Platysma.	G. *platysma*, plate.
Plexus (pl. **-us**).	L. *plexus*, plaiting, braid.
Pollex.	L. *pollex*, thumb.
Pronator.	L. *pronare*, to turn face downward.

Protuberance.	L. *protubero*, I swell.
Proximal.	L. *proximus*, next.
Quadrangular.	L. *quattuor* four + L. *angulus*, four angles.
Radial.	L. *radius*, rod, spoke.
Radicle.	L. *radix*, root.
Radius.	L. *radius*, rod, spoke.
Radix.	L. *radix*, root.
Ramus.	L. *ramus*, branch.
Raphe.	G. *raphe*, seam.
Reflect.	L. *reflecto*, to turn back.
Retraction.	L. *re*, back, + *trahere*, to draw.
Retractor.	L. *retrahere*, to draw back.
Rotator.	L. *rotare*, to whirl about.
Sagittal.	L. *sagitta*, arrow; shape of saggital suture including the lambdoid suture.
Scalene.	G. *skalenos*, uneven.
Scaphoid.	G. *skaphe*, boat, + *eidos*, shape, likeness.
Scapula.	L. *scapulae*, shoulder-blades.
Sesamoid.	G. *sesamen*, sesame plant, or seed, + *eidos*, shape, likeness.
Sinister.	L. *sinister*, left side or unlucky.
Somatic.	G. *soma* (pl. *somata*), body.
Somite.	G. *soma*, body, + suffix -*ite*, indicating origin.
Spina.	L. *spina*, thorn.
Sternum.	G. *sternon*, chest.
Supination.	L. *supinus*, lying on the back.
Supinator.	L. *supinare*, to bend backward.
Supraspinatus.	L. *supra*, above, + *spina*, thorn.
Synapse.	G. *syn*, together, + *haptein*, to fasten.
Synarthrosis.	G. *syn*, together, + *arthron*, joint.

Synchrondosis.	G. *syn*, together, + *chondros*, cartilage.
Syndesmosis.	G. *syn*, together, + *desmosis*, band.
Synergy.	G. *syn*, together, + *ergon*, work.
Synostosis.	G. *syn*, together, + *osteon*, bone.
Synovia.	G. *syn*, together, + *ovum*, egg.
Tendon.	L. *tendere*, to stretch.
Tensor.	L. *tendere*, to stretch.
Teres.	L. *tero*, I grind, rub.
Theca.	L. *theca*, envelope, sheath.
Thenar.	G. *thenar*, hand.
Trapezius.	G. *trapezion*, small four-sided table.
Triceps.	L. *tres*, three, + *caput*, head.
Triquetrum.	L. *triquetrus*, three-cornered, triangular.
Trochanter.	G. *trochos*, wheel, pulley.
Trochlea.	G. *trochilia*, pulley.
Tuberosity.	L. *tuber*, round smooth swelling.
Ulna.	L. *ulna*, elbow.
Uncus.	L. *uncus*, hook.
Unipennate.	L. *unus*, one, + *penna*, feather.
Ventral.	L. *venter*, belly.
Vestige.	L. *vestigium*, trace, footprint.
Volar.	L. *vola*, palm of hand.
Xiphoid.	G. *xiphos*, sword, + *eidos*, shape, likeness.

TERMS IMPORTANT IN THE LOWER LIMB

Abductor.	L. *ab*, away, + *ducere*, to lead.
Acetabulum.	L. *acetabulum*, vinegar cup, shape of hip joint.
Adductor.	L. *ad*, to, + *ducere*, to lead.
Adhesion.	L. *adhaereo*, to stick together.
Adipose.	L. *adeps*, fat.
Amphiarthrosis.	G. *amphi*, on both sides, + *arthrosis*, joint.
Anastomosis.	G. *anastomoein*, to bring to a mouth, cause to communicate.
Ankle.	L. *angulus*, angle.
Annulus.	L. *anulus* (*annulus*), a ring.
Anterior.	L. *ante*, before.
Aponeurosis.	G. *apo*, from, + *neuron*, tendon (earlier: nerve, later: tendon).
Arcuate.	L. *arcualis*, arch-shaped.
Artery.	G. *aer*, air, + *terein*, to keep (L. *arteria*, windpipe); ancient belief that blood vessels contained air.
Articulation.	L. *artus*, joint, *articulatus*, little joint, pl. *articulationes*, joints.
Aspera.	L. *asper*, rough.
Axial.	L. *axis*, axle of a wheel, the line about which any body turns.
Axis.	L. *axis*, axle of a wheel.

Biceps.	L. *bis*, two, + *caput*, head.
Bilateral.	L. *bi*, two, + *latus*, side.
Bipennate.	L. *bis*, two, + *pinna*, feather.
Brevis.	L. *brevis*, short.
Bursa.	L. *bursa*, a purse; hence purse-shaped object.
Calcaneus.	L. *calcaneus*, heel.
Calcar.	L. *calcar*, spur.
Calcification.	L. *calx*, lime, + *facere*, to make.
Capsule.	L. dim. of *capsa*, box.
Caput (pl. capita).	L. *caput*, head.
Cauda, caudal.	L. *cauda*, tail.
Cingulum.	L. *cingulum*, girdle.
Circumflex.	L. *circum*, around, + *flexere*, to bend.
Collateral.	L. *con*, together, + *latus*, side.
Comes.	L. *comes*, companion.
Comitans (pl. comitantes).	L. *comitari*, to accompany.
Commissure.	L. *commissura* (*cum* + *mittere*), connection.
Communicans.	L. *communicans*, communicating.
Condyle.	G. *kondylos*, knuckle.
Coxa.	L. *coxa*, hip.
Crista.	L. *crista*, crest.
Cruciate.	L. *crux*, cross.
Crus (pl. crura).	L. *crus*, leg.
Cuboid.	G. *kuboeides*, cube-shaped.
Cuneiform.	L. *cuneus*, wedge, + *forma*, shape, likeness.
Diaphysis.	G. *dia*, between, + *physis*, growth.
Diarthrosis.	G. *dia*, through, + *arthroun*, to fasten by a joint.
Digit.	L. *digitus*, finger.

Disc.	L. *discus*, disc.
Distal.	L. *distare*, to stand apart.
Dorsal.	L. *dorsum*, back.
Eminence.	L. *e*, out, + *minere*, to jut.
Emissary.	L. *e*, out, + *mittere*, to send.
Epicondyle.	G. *epi*, on, + *kondylos*, knuckle.
Epiphysis.	G. *epi*, on, + *physis*, growth.
Eversion.	L. *e*, out, + *vertere*, to turn.
Extension.	L. *extendo*, extend.
Extensor.	L. *ex*, out, + *tendere*, to stretch.
Extrinsic.	L. *extrinsecus*, on the outside.
Falciform.	L. *falx*, sickle, + *forma*, shape, likeness.
Fascia (pl. -iae).	L. *fascia*, band.
Fasciculus.	L. *fascis*, bundle.
Femur.	L. *femur*, thigh.
Fibula.	L. *fibula*, pin, skewer, brooch.
Flexor.	L. *flexus*, bent.
Foramen.	L. *foramen*, hole.
Fossa.	L. *fossa*, ditch, channel, something dug.
Fulcrum.	L. *fulcrum*, post.
Fusiform.	L. *fusus*, spindle, + *forma*, shape, likeness.
Gastrocnemius.	G. *gaster*, belly, + *kneme*, leg.
Gemellus.	L. *gemellus*, twin.
Geniculate.	L. *geniculatus*, with bent knee.
Genu.	L. *genu*, knee.
Ginglymus.	G. *ginglymus*, hinge.
Gluteus.	G. *gloutos*, rump.
Gracilis.	L. *gracilis*, thin.

Hernia.	L. *hernia*, protrusion.
Hiatus.	L. *hiatus*, gap.
Hilum.	L. *hilum*, a small thing.
Hypaxial.	G. *hypo*, under, + L. *axis*, centre line, axis.
Iliacus.	See *ilium*.
Ilium.	L. *ileum*, flank.
Incisura.	L. *incidere*, to cut into.
Inductor.	L. *inducere*, to lead on, excite.
Inguinal.	L. *inguina*, groin.
Innominate.	L. *in*, in, + *nomen*, name, unnamed bone, unnamed artery.
Insertion.	L. *in*, in, + *serere*, to plant.
Intrinsic.	L. *intrinsecus*, inward.
Invagination.	L. *in*, in, + *vagina*, sheath.
Inversion.	L. *in*, in, + *vertere*, to turn.
Ischiofemoral.	G. *ischion*, hip, + L. *femur*, thigh.
Ischium (pl. -ia).	G. *ischion*, hip.
Kinetic.	G. *kinesis*, movement.
Labrum.	L. *labrum*, rim.
Lateral.	L. *lateralis* (*latus*), pertaining to a side.
Ligamentum.	L. *ligamentum*, a bandage.
Locus.	L. *locus*, place.
Longus.	L. *longus*, long.
Lumbrical.	L. *lumbricus*, earthworm.
Magnus, -a, -um.	L. *magnus*, large.
Marginal.	L. *marginalis* (*margo*), bordering.
Medial.	L. *medialis* (*medius*), pertaining to the middle.

Median.	L. *medianus*, in the middle.
Meniscus.	L. *menis*, cresent, half-moon, dim. of *mene*, moon.
Metatarsus.	G. *meta*, after, + L. *tarsus*, ankle.
Minimus.	L. *minimus*, least.
Nates.	L. *nates*, buttocks.
Navicular.	L. *navicula*, small boat.
Obliquus.	L. *obliquus*, slanting.
Obturator.	L. *obturo*, I stop up.
Os.	L. *os*, bone.
Osseous.	L. *os* (pl. *ossa*), bone.
Ossicle.	L. *ossiculum*, small bone.
Ossification.	L. *os*, bone, + *facere*, to make.
Osteone.	G. *osteon*, bone.
Osteomalacia.	G. *osteon*, bone, + *malakia*, softness.
Paraxial.	G. *para*, alongside of, + L. *axis*, axle.
Patella.	L. *patella*, small pan.
Pedicle.	L. *pes*, foot.
Pelvic.	L. *pelvis*, basin.
Penniform.	L. *penna*, feather, + *forma*, form, likeness.
Peroneal.	G. *perone*, = L. *fibula*, pin; hence pertaining to needle-shaped leg bone.
Pes.	L. *pes*, foot.
Phalanges.	G. *phalanx*, band of soldiers; singular phalanx, originally denoted whole set of bones of a digit not just one bone.
Piriform.	L. *pirum*, pear, + *forma*, shape, likeness.
Planta.	L. *planta*, sole of the foot.
Plantigrade.	L. *planta*, side of foot, + *gradior*, to walk.

Platycnemia.	G. *platys*, flat, + *kneme*, knee; hence condition of side-to-side flattening of tibia giving prominence to it's anterior border.
Platymeria.	G. *platys*, flat, + *meros*, thigh; as previous for thigh.
Popliteus.	L. *poples*, ham.
Profundus.	L. *profundus*, deep.
Promontory.	G. *promontorium*, mountain ridge.
Pronator.	L. *pronare*, to turn face downward.
Pronograde.	L. *pronus*, bent downward, + *gradus* step.
Protuberance.	L. *protubero*, I swell.
Proximal.	L. *proximus*, next.
Psoas.	G. *psoa*, loin.
Pubes.	L. *pubes*, mature.
Pubis (pl. -es).	L. *pubes*, mature.
Pyriform.	L. *pirum*, pear, + *forma*, shape.
Quadrangular.	L. *quattuor* four + L. *angulus*, four angles.
Quadriceps.	L. *quattuor*, four, + *caput*, head.
Raphe.	G. *raphe*, seam.
Reflect.	L. *reflecto*, to turn back.
Retraction.	L. *re*, back, + *trahere*, to draw.
Retractor.	L. *retrahere*, to draw back.
Rotator.	L. *rotare*, to whirl about.
Saphenous.	A. *al-safin*, hidden (later G. *saphenes*, visible).
Sartorius.	L. *sartor*, tailor.
Sciatic.	G. *ischion*, hip joint.
Semilunar.	L. *semi*, half, + *luna*, moon.
Semimembranosus.	L. *semi*, half, + *membranosus*, membrane.
Semitendinosus.	L. *semi*, half, + *tendinosus*, tendon.

Sesamoid.	G. *sesamen*, sesame plant, or seed, + *eidos*, shape, likeness.
Sinister.	L. *sinister*, left side or unlucky.
Soleus.	L. *solea*, sandal, sole.
Spina.	L. *spina*, thorn.
Supination.	L. *supinus*, lying on the back.
Supinator.	L. *supinare*, to bend backward.
Sural.	L. *sura*, calf of leg.
Sustentaculum.	L. *sustentaculum*, support.
Synovia.	G. *syn*, together, + *ovum*, egg.
Talipes.	L. *talipedo*, weak in the feet (*talipes*, clubfoot).
Talonid.	L. *talus*, ankle bone, + G. *eidos*, form, likeness.
Talus.	L. *talus*, ankle-bone.
Tarsus.	G. *tarsos*, sole of the foot.
Tendon.	L. *tendere*, to stretch.
Tensor.	L. *tendere*, to stretch.
Teres.	L. *tero*, I grind, rub.
Tibia.	L. *tibia*, long flute.
Tuberosity.	L. *tuber*, round smooth swelling.
Ungual.	L. *unguis*, claw, nail.
Unipennate.	L. *unus*, one, + *penna*, feather.
Valgus.	L. *valgus*, bow-legged (**Genu valgus** now means knock-kneed).
Varus.	L. *varus*, knock-kneed (term has become transposed, see **Valgus**).
Ventral.	L. *venter*, belly.
Vincula.	L. *vinculum*, band, cord.

TERMS IMPORTANT IN THE THORAX

Accessory.	L. *accessorius*, supplementary.
Acinus.	L. *acinus*, grape.
Adhesion.	L. *adhaereo*, to stick together.
Adipose.	L. *adeps*, fat.
Aditus.	L. *aditus*, opening.
Adnexa.	L. *ad*, to, + *nexus*, bound.
Adventia.	L. *ad*, to, + *venire*, to come.
Afferent.	L. *ad*, to, + *ferre*, to carry.
Alveolus.	L. *alveolus*, little cavity.
Annulus.	L. *anulus* (*annulus*), a ring.
Anterior.	L. *ante*, before.
Atrium.	L. *atrium*, court, entrance hall.
Auricle.	L. dim., *auricula*, external ear, also shape of heart chamber.
Autonomic.	G. *autos*, self, + *nomos*, law.
Axon.	G. *axon*, axis.
Azygos.	G. *a*, not, + *zygon*, yoke; hence, unpaired.
Bolus.	G. *bolos*, mass.
Bronchiole.	G. *bronchiolus* (dim. of *bronchus*, *brechein*, to moisten).
Bronchus.	G. *bronchia*, end of windpipe.
Bulbus.	L. *bulbus*, bulb, swollen root.

Bursa.	L. *bursa*, a purse; hence purse-shaped object.
Cardiac.	G. *kardiakos* (*kardia*), pertaining to the heart.
Carina.	L. *carina*, keel.
Carotid.	G. *karon*, deep sleep (pressure on artery produces stupor).
Cartilage.	L. *cartilago*, gristle, cartilage.
Cava.	L. *cavus, -a, -um*, hollow or cave.
Centrum.	L. *centrum*, centre.
Chorda.	G. *chorde*, string of gut, cord.
Circumflex.	L. *circum*, around, + *flexere*, to bend.
Cisterna.	L. *cisterna*, reservoir.
Collateral.	L. *con*, together, + *latus*, side.
Comes.	L. *comes*, companion.
Comitans (pl. **comitantes**).	L. *comitari*, to accompany.
Communicans.	L. *communicans*, communicating.
Conus.	L. *conus*, cone.
Cornu.	L. *cornu*, horn.
Corona.	L. *corona*, crown.
Coronary.	L. *coronarius*, pertaining to a wreath or crown; hence, encircling.
Coronoid.	G. *korax*, crow, + *eidos*, form, likeness.
Costa (adj. **costal**).	L. *costa*, rib.
Crista.	L. *crista*, crest.
Crus (pl. **crura**).	L. *crus*, leg.
Deglutition.	L. *deglutire*, to swallow.
Dermatome.	G. *derma*, skin, + *temnein*, to cut.
Diaphragm.	G. *dia*, through, + *phragma*, wall.
Diastole.	G. *dia*, through, + *stellein*, to send.
Disc.	L. *discus*, disc.

Diverticulum.	L. *divertere*, to turn aside.
Dorsal.	L. *dorsum*, back.
Duct.	L. *ducere*, to lead or draw.
Ductule.	L. *ducere*, to lead, dim.
Efferent.	L. *ex*, out, + *ferre*, to bear.
Embolus.	G. *embolos*, wedge, plug, anything inserted.
Eminence.	L. *e*, out, + *minere*, to jut.
Emissary.	L. *e*, out, + *mittere*, to send.
Endocardium.	G. *endon*, within, + *kardia*, heart.
Endochondral.	G. *endon*, within, + *chondros*, cartilage.
Endomysium.	G. *endon*, within, + *mys*, muscle.
Endoneurium.	G. *endon*, within, + *neuron*, nerve.
Endosteum.	G. *endon*, within, + *osteon*, bone.
Endothelium.	G. *endon*, within, + *thele*, nipple.
Epicardium.	G. *epi*, on, + *kardia*, heart.
Epidermis.	G. *epi*, on, + *derma*, skin.
Epigastrium.	G. *epi*, upon, + *gaster*, belly.
Epimysium.	G. *epi*, upon, + *mys*, muscle.
Epineurium.	G. *epi*, on, + *neuron*, nerve.
Epiphysis.	G. *epi*, on, + *physis*, growth.
Extension.	L. *extendo*, extend.
Extensor.	L. *ex*, out, + *tendere*, to stretch.
Extravasation.	L. *extra*, outside, + *vas*, vessel.
Extrinsic.	L. *extrinsecus*, on the outside.
Fascia (pl. -iae).	L. *fascia*, band.
Fenestra.	L. *fenestra*, window.
Filament.	L. *filamentum*, thin fibre.
Fistula.	L. *fistula*, pipe.
Flaccid.	L. *flaccidus*, weak.

Flexor.	L. *flexus*, bent.
Foramen.	L. *foramen*, hole.
Gland.	L. *glandula*, dim. *glans*, acorn, pellet.
Hernia.	L. *hernia*, protrusion.
Hiatus.	L. *hiatus*, gap.
Hilum.	L. *hilum*, a small thing.
Hypaxial.	G. *hypo*, under, + L. *axis*, centre line, axis.
Incisura.	L. *incidere*, to cut into.
Innervation.	L. *in*, in, + *nervus*, nerve.
Insertion.	L. *in*, in, + *serere*, to plant.
Integument.	L. *in*, over, + *tegere*, to cover.
Intercalated.	L. *inter*, between, + *calare*, to call, inserted or placed between.
Intercostal.	L. *inter*, between, + *costa*, rib.
Interdigitating.	L. *inter*, between, + *digitus*, digit, interlocked by finger-like processes.
Intervertebral.	L. *inter*, between, + *vertebra*, joint.
Intima.	L. *intima*, innermost.
Intrinsic.	L. *intrinsecus*, inward.
Invagination.	L. *in*, in, + *vagina*, sheath.
Inversion.	L. *in*, in, + *vertere*, to turn.
Isthmus.	G. *isthmos*, narrow connection.
Juxtaposition.	L. *juxta*, near, + *positio*, place.
Kyphosis.	G. *kyphos*, bent.
Lacerate.	L. *lacerare*, to tear.
Lactation.	L. *lactare*, to suckle.
Lamella.	L. dim. of *lamina*, leaf.

Lamina (pl. -ae).	L. *lamina*, thin plate.
Lateral.	L. *lateralis* (*latus*), pertaining to a side.
Levator.	L. *levare*, to raise.
Ligamentum.	L. *ligamentum*, a bandage.
Lingula.	L. *lingula*, small tongue.
Lobus.	G. *lobos*, lobe.
Locus.	L. *locus*, place.
Lordosis.	G. *lordoo*, I bend.
Lumen.	L. *lumen*, light, opening.
Lung.	AS. *lunge*, lung.
Mammary.	L. *mamma*, breast.
Mammillary.	L. dim. *mammillaris* (*mamma*, -ae), little breast.
Manubrium.	L. *manubrium*, handle, hilt (as of a sword).
Marginal.	L. *marginalis* (*margo*), bordering.
Medial.	L. *medialis* (*medius*), pertaining to the middle.
Median.	L. *medianus*, in the middle.
Mediastinum.	L. *mediastinus*, servant, drudge, but anatomical term mediastinum probably derived from *per medium tensum,* tight in the middle.
Membrane.	L. *membrana*, skin.
Mesothelium.	G. *mesos*, middle, + *thele*, nipple, hence middle lining layer.
Mitral.	L. *mitra*, turban, but Hebrew priest headress the only head covering that the mitral valve resembles.
Muscle.	L. *musculus*, little mouse.
Myocardium.	G. *mys, myos*, muscle, + *kardia*, heart.
Myology.	G. *mys*, muscle, + *logos*, treatise.

Nerve.	L. *nervus*, G. *neuron*, cordlike structure, nerve, tendon.
Neural.	G. *neuron*, nerve.
Neuralgia.	G. *neuron*, nerve, + *algos*, pain.
Neuraxon.	G. *neuron*, nerve, + *axon*, axis.
Neurilemma.	G. *neuron*, nerve, + *lemma*, husk, sheath.
Neuroblast.	G. *neuron*, nerve, + *blastos*, bud.
Neuroectomy.	G. *neuron*, nerve, + *ektome*, excision.
Neuroglia.	G. *neuron*, nerve, + *gloia*, glue.
Neurology.	G. *neuron*, nerve, + *logos*, treatise.
Neuron.	G. *neuron*, cordlike structure, sinew, tendon; equivalent of L. *nervus,* hence, nerve.
Node.	L. *nodus*, knot.
Oedema, Edema.	G. *oidema*, swelling.
Oesophagus, Esophagus.	G. *oisein* (*phero*), to carry, + *phagein*, to eat.
Organ.	L. *organum*, implement.
Orifice.	L. *orificium*, opening.
Os.	L. *os*, bone.
Osseous.	L. *os* (pl. *ossa*), bone.
Ossicle.	L. *ossiculum*, small bone.
Ossification.	L. *os*, bone, + *facere*, to make.
Ostium.	L. *ostium*, door.
Paraganglion.	G. *para*, along, beside, + *ganglion*, knot.
Paramedian.	G. *para*, alongside of, + *mesos*, middle.
Parasternal.	G. *para*, alongside of, + *sternon*, chest.
Parasympathetic.	G. *para*, alongside of, + *sympathetikos*, sympathetic.
Paraxial.	G. *para*, alongside of, + L. *axis*, axle.
Parenchyma.	G. *parenchyma*, pouring out into the adjacent.
Parietal.	L. *paries*, wall.

Parieties.	L. *paries*, wall.
Pectoral.	L. *pectoralis* (*pectus*), belonging to the breast.
Percussion.	L. *percussio*, striking.
Pericardial.	G. *peri*, around, + *kardia*, heart.
Perichondrium.	G. *peri*, around, + *chondros*, cartilage.
Perilymph.	G. *peri*, around, + L. *lympha*, fluid.
Perimysium.	G. *peri*, around, + *mys*, muscle.
Periosteum.	G. *peri*, around, + *osteon*, bone.
Peristalsis.	G. *peristaltikos*, clasping and compressing.
Pleura.	G. *pleura*, rib; ie related to the ribs.
Plexus (pl. -us).	L. *plexus*, plaiting, braid.
Pneumatic.	G. *pneumatikos*, pertaining to breath.
Praecordium, Precordium.	L. *prae*, in front of, + *cordis*, of the heart.
Proximal.	L. *proximus*, next.
Pulmonary.	L. *pulmo*, lung.
Pulmones.	L. *pulmo*, lung.
Radicle.	L. *radix*, root.
Radix.	L. *radix*, root.
Ramify.	L. *ramus*, branch, + *facere*, to make.
Ramus.	L. *ramus*, branch.
Raphe.	G. *raphe*, seam.
Recess.	L. *recessus*, retreat.
Reflect.	L. *reflecto*, to turn back.
Sac.	L. *saccus*, sack.
Sacculus.	L. *sacculus*, a little bag.
Sarcolemma.	G. *sarx*, flesh, + *lemma*, husk, skin.
Septum.	L. *saeptum*, fence.
Serratus.	L. *serra*, saw.
Sinister.	L. *sinister*, left side or unlucky.

Sinus (pl. -us).	L. *sinus*, curve, cavity, bosom.
Sinusoid.	L. *sinus*, curve, cavity, + *eidos*, shape, likeness.
Situs inversus viscerum.	L. *situs*, site, position, + *inversus*, inverted, + *viscerum*, of the viscera.
Somatic.	G. *soma* (pl. *somata*), body.
Somatopleure.	G. *soma*, body, + *pleura*, side.
Somite.	G. *soma*, body, + suffix *-ite*, indicating origin.
Sphincter.	G. *sphingein*, to bind tight.
Splanchnic.	G. *splanchna*, viscera.
Stenosis.	G. *stenosis*, narrowing.
Sternum.	G. *sternon*, chest.
Subcostal.	L. *sub*, under, + *costa*, rib.
Supracostal.	L. *supra*, above, + *costa*, rib.
Sympathetic.	G. *syn*, together, + *pathein*, to suffer.
Synapse.	G. *syn*, together, + *haptein*, to fasten.
Syncytium.	G. *syn*, together, + *cytos*, cell.
Syndrome.	G. *syndrome*, occurrence.
Systole.	G. *systole*, contraction.
Theca.	L. *theca*, envelope, sheath.
Thorax (pl. thoraces).	G. *thorax*, breast-plate, breast.
Thrombus.	G. *thrombus*, lump.
Thymus.	G. *thymos*, thyme.
Trachea.	G. *tracheia*, windpipe.
Truncus.	L. *truncus*, trunk of tree.
Tunica.	L. *tunica*, undergarment.
Vagus.	L. *vagus*, wandering.
Valve.	L. *valva*, leaf of door.
Valvula.	L. *valvula*, a little fold, valve.
Vas.	L. *vas*, vessel.

Vascular.	L. *vasculum*, small vessel.
Vein.	L. *vena*, vein.
Velum.	L. *velum*, veil.
Ventral.	L. *venter*, belly.
Ventricle.	L. *ventriculus*, little cavity, loculus.
Vertebra.	L. *vertebra*, joint.
Vertex.	L. *vertex*, whirl, whirlpool.
Vincula.	L. *vinculum*, band, cord.
Visceral.	L. *viscera*, entrails, bowels.
Viscus.	L. *viscus*, internal organ.
Xiphoid.	G. *xiphos*, sword, + *eidos*, shape, likeness.
Zygapophysis.	G. *zygon*, yoke, + *apophysis*, process of a bone.

TERMS IMPORTANT IN THE ABDOMEN AND PELVIS

Abdomen.	L. from *abdere* (?), to hide.
Aberrant.	L. *ab*, away, + *errare*, to stray.
Abortion.	L. *abortio*, to abort.
Absorption.	L. *absorptio*, to swallow.
Accessory.	L. *accessorius*, supplementary.
Acinus.	L. *acinus*, grape.
Adhesion.	L. *adhaereo*, to stick together.
Adipose.	L. *adeps*, fat.
Aditus.	L. *aditus*, opening.
Adnexa.	L. *ad*, to, + *nexus*, bound.
Adrenal.	L. *ad*, near, + *renes*, kidneys.
Adventia.	L. *ad*, to, + *venire*, to come.
Afferent.	L. *ad*, to, + *ferre*, to carry.
Allantois.	G. *allas*, sausage, + *eidos*, form, appearance.
Alveolus.	L. *alveolus*, little cavity.
Amnion.	G. *amnion*, fetal membrane.
Ampulla.	L. *ampulla*, a flask or vessel swelling in the middle.
Anastomosis.	G. *anastomoein*, to bring to a mouth, cause to communicate.
Anlage.	Ger. *an*, on, + *legen*, to lay (a laying on — *primordium*, precursor).
Annulus.	L. *anulus* (*annulus*), a ring.
Anterior.	L. *ante*, before.

Anteversion.	L. *ante*, before, + *versio*, turning.
Antrum.	G. *antron*, cave.
Anus.	L. *anus*, fundament.
Aorta.	G. *aeiro*, to raise.
Apertura.	L. *apertura*, opening.
Appendicular.	adjectival form of appendix.
Appendix.	L. *appendere*, to hang upon.
Arcuate.	L. *arcualis*, arch-shaped.
Arcus.	L. *arcus*, bow.
Atresia.	G. *a*, without, + *tresis*, hole.
Atrium.	L. *atrium*, court, entrance hall.
Atrophy.	G. *a*, without, + *trophe*, nourishment.
Autonomic.	G. *autos*, self, + *nomos*, law.
Axial.	L. *axis*, axle of a wheel, the line about which any body turns.
Bicuspid.	L. *bis*, two, + *cuspis*, point.
Bilateral.	L. *bi*, two, + *latus*, side.
Bile.	L. *bilis*, bile.
Biventer.	L. *bis*, two, + *venter*, belly.
Bolus.	G. *bolos*, mass.
Calyx (pl. calices).	L. *calyx*, husk, cup-shaped protective covering.
Cauda, caudal.	L. *cauda*, tail.
Cava.	L. *cavus*, -*a*, -*um*, hollow or cave.
Cavernosus.	L. *caverna*, a hollow or cave.
Cavum.	L. *cavum*, a hollow or cave.
Cecum, Caecum.	L. *caecus*, -*a*, -*um*, blind.
Celiac, Coeliac.	G. *koilia*, belly.
Celom (e), Coelom.	G. *koiloma*, a hollow.
Centrum.	L. *centrum*, centre.

Choledochus.	G. *chole*, bile, + *dochos* (*dechomai*), container.
Chyle.	G. *chylos*, juice.
Chyme.	G. *chymos*, juice.
Circumflex.	L. *circum*, around, + *flexere*, to bend.
Cisterna.	L. *cisterna*, reservoir.
Cloaca.	L. *cloaca*, sewer, drain.
Coccyx.	G. *kokkyx*, a cuckoo, hence a structure shaped like a cuckoo's bill.
Colon.	G. *kolon*, great gut.
Comes.	L. *comes*, companion.
Comitans (pl. **comitantes**).	L. *comitari*, to accompany.
Commissure.	L. *commissura* (*cum* + *mittere*), connection.
Communicans.	L. *communicans*, communicating.
Coraco-, corono-.	G. *korax* or *corone*, crow; hence crowlike.
Cornu.	L. *cornu*, horn.
Corona.	L. *corona*, crown.
Corpus (pl. **corpora**).	L. body.
Cremaster.	G. *cremaster*, suspender.
Crus (pl. **crura**).	L. *crus*, leg.
Cyst (adj. **cystic**).	G. *kystis*, bag, bladder, pouch.
Dartos.	G. *dartos*, skinned.
Decidua, (adj. **Deciduous**).	L. *deciduus*, (*de* + *cado*), falling off.
Defaecation, Defecation.	L. *defaecare*, to cleanse.
Deferens.	L. *de*, away, + *ferens*, carrying.
Detritus.	L. *deterere*, to rub away.
Detrusor.	L. *detrudere*, to push down.
Diaphragm.	G. *dia*, through, + *phragma*, wall.

Digastric.	G. *dis*, double, + *gaster*, belly.
Digestion.	L. *dis*, apart, + *gerere*, to carry.
Disc.	L. *discus*, disc.
Distal.	L. *distare*, to stand apart.
Diverticulum.	L. *divertere*, to turn aside.
Dorsal.	L. *dorsum*, back.
Duct.	L. *ducere*, to lead or draw.
Ductule.	L. *ducere*, to lead, dim.
Duodenum.	L. *duodeni*, twelve (meaning twelve fingerbreadths).

Ejaculatory.	L. *e*, out, + *jacere*, to throw.
Emboliformis.	G. *embolos*, wedge, + L. *forma*, shape.
Embolus.	G. *embolos*, wedge, plug, anything inserted.
Endometrium.	G. *endon*, within, + *metra*, womb.
Endothelium.	G. *endon*, within, + *thele*, nipple.
Epididymis.	G. *epi*, on, + *didmoi*, testicles.
Epigastrium.	G. *epi*, upon, + *gaster*, belly.
Epiploic.	G. *epiploon*, caul, omentum.
Epispadias.	G. *epi*, upon, + N.L. *spadias*, F. *spadon*, eunuch, F. *span*, to draw

Erector.	L. *erectus*, erect.
Exomphalos.	G. *ex*, out, + *omphalos*, navel.
Excretion.	L. (*excretus*) *ex*, out, + *cernere*, to sift.
Extravasation.	L. *extra*, outside, + *vas*, vessel.
Extrinsic.	L. *extrinsecus*, on the outside.
Exudate.	L. *ex*, out, + *sudare*, to sweat.

Faeces, Feces.	L. *faex*, dregs.
Falciform.	L. *falx*, sickle, + *forma*, shape, likeness.
Falx.	L. *falx, falcis*, sickle.
Fascia (pl. -iae).	L. *fascia*, band.
Fasciculus.	L. *fascis*, bundle.

Fenestra.	L. *fenestra*, window.
Fetus, Foetus.	L. *fetus*, offspring.
Filum.	L. *filum*, thread.
Fimbria.	L. *fimbriae*, threads, fringe.
Fissure.	L. *fissura* (*findo*), a cleft.
Fistula.	L. *fistula*, pipe.
Flaccid.	L. *flaccidus*, weak.
Follicle.	L. *folliculus*, a small bag.
Foramen.	L. *foramen*, hole.
Fornix.	L. *fornix*, arch or vault.
Fossa.	L. *fossa*, ditch, channel, something dug.
Fourchette.	F. *fourchette*, fork.
Fraenum, Frenum.	L. *fraenum*, bridge.
Frenulum.	L. dim. *fraenum*, a little bridge.
Fundiform.	L. *funda*, sling, + *forma*, shape, likeness.
Fundus.	L. *fundus*, bottom.
Fusiform.	L. *fusus*, spindle, + *forma*, shape, likeness.
Genital.	L. *genitalis* (*gigno*), pertaining to birth.
Germinal.	L. *germen*, bud, germ.
Germinative.	L. *germen*, bud, germ.
Gestation.	L. *gestare*, to bear.
Gland.	L. *glandula*, dim. *glans*, acorn, pellet.
Glans.	L. *glans*, acorn.
Glomerulus.	L. dim. of *glomus*, a ball.
Gonad.	G. *gone*, seed.
Granulation.	L. dim. *granum*, grain.
Granulum.	L. *granulum*, small grain.
Gravid.	L. *gravida*, pregnant.
Gubernaculum.	L. *gubernaculum*, helm.

Haemal, hemal.	G. *haima*, blood.
Hemoglobin,	G. *haima*, blood, + L. *globus*, sphere.
Haemoglobin.	
Hemopoietic,	G. *haima*, blood, + *poietikos*, creative.
Haemopoietic.	
Hepar.	G. *hepar*, liver.
Hepatic.	L. *hepar*, liver.
Hernia.	L. *hernia*, protrusion.
Hiatus.	L. *hiatus*, gap.
Hilum.	L. *hilum*, a small thing.
Hydrocoele,	G. *hydor*, water, + *koilos*, hollow.
Hydrocele.	
Hymen.	G. *hymen*, membrane.
Hypaxial.	G. *hypo*, under, + L. *axis*, centre line, axis.
Hypospadias.	G. *hypo*, under, + N.L. *spadias*, F. *spadon*, F. *span*, to draw.

Icterus.	G. *ikteros*, jaundice.
Ileum.	G. *eilein*, to wind or turn.
Implantation.	L. *in*, in, + *plantere*, to plant.
Impression.	L. *in*, in, + *premere*, to press.
Incisura.	L. *incidere*, to cut into.
Inguinal.	L. *inguina*, groin.
Innervation.	L. *in*, in, + *nervus*, nerve.
Intercostal.	L. *inter*, between, + *costa*, rib.
Interdigitating.	L. *inter*, between, + *digitus*, digit, interlocked by finger-like processes.
Intervertebral.	L. *inter*, between, + *vertebra*, joint.
Intestine.	L. *intus*, within, or L. *intestina*, guts, entrails.
Intima.	L. *intima*, innermost.
Intrinsic.	L. *intrinsecus*, inward.
Introitus.	L. *intro*, within, + *ire*, to go.
Invagination.	L. *in*, in, + *vagina*, sheath.

Inversion.	L. *in*, in, + *vertere*, to turn.
Isthmus.	G. *isthmos*, narrow connection.
Jejunum.	L. *jejunus*, empty (of food), that part of intestine that appears empty.
Juxtaposition.	L. *juxta*, near, + *positio*, place.
Labial.	L. *labialis* (*labia*), pertaining to lips.
Labium.	L. *labium*, lip.
Labrum.	L. *labrum*, rim.
Lacteals.	L. *lactare*, to suckle.
Laparotomy.	L. *lapara*, flank, + *tome*, section, cut.
Lien.	L. *lien*, spleen.
Lieno-.	L. *lien*, spleen from original L. *splen* (the sp having been dropped).
Ligamentum.	L. *ligamentum*, a bandage.
Ligature.	L. *ligare*, to bind.
Liver.	AS. *lifer*, liver.
Lobus.	G. *lobos*, lobe.
Locus.	L. *locus*, place.
Lordosis.	G. *lordoo*, I bend.
Lumbar.	L. *lumbare*, apron for the loins.
Lumen.	L. *lumen*, light, opening.
Lung.	AS. *lunge*, lung.
Luteum.	L. *luteus*, yellow.
Marginal.	L. *marginalis* (*margo*), bordering.
Matrix.	L. *matrix* (*mater*), womb, groundwork, mold.
Meatus (pl. -us).	L. *meatus*, passage.
Medial.	L. *medialis* (*medius*), pertaining to the middle.
Median.	L. *medianus*, in the middle.
Membrane.	L. *membrana*, skin.
Menopause.	G. *men*, month, + *pausis*, cessation.

Menstruation.	G. *men*, month, L. *menstruus*, pertaining to a month.
Mesenchyme.	G. *mesos*, middle, + *en*, in, + *chymos*, juice.
Mesentery.	G. *mesos*, midway between, + *enteron*, gut.
Mesocolon.	G. *mesos*, middle, + *kolon*, great gut.
Mesogastrium.	G. *mesos*, middle, + *gaster*, stomach.
Mesonephros.	G. *mesos*, middle, + *nephros*, kidney.
Mesosalpinx.	G. *mesos*, middle, + *salpinx*, tube.
Mesothelium.	G. *mesos*, middle, + *thele*, nipple, hence middle lining layer.
Mesovarium.	G. *mesos*, middle, + L. *ovarium*, ovary.
Metanephros.	G. *meta*, after, + *nephros*, kidney.
Mucus.	L. *mucus*, G. *muxa*, snivel, slippery secretion.
Multiparous.	L. *multus*, many, + *parire*, to give birth.
Myenteric.	G. *mys*, muscle, + *enteron*, gut.
Myocardium.	G. *mys*, *myos*, muscle, + *kardia*, heart.
Navel.	AS. *nafe*, centre of hub of wheel.
Necrosis.	G. *nekrosis*, a killing.
Necropsy.	G. *nekros*, corpse, + *opsis*, sight.
Nerve.	L. *nervus*, G. *neuron*, cordlike structure, nerve, tendon.
Neural.	G. *neuron*, nerve.
Neuralgia.	G. *neuron*, nerve, + *algos*, pain.
Neuraxon.	G. *neuron*, nerve, + *axon*, axis.
Neurenteric.	G. *neuron*, nerve, + *enteron*, gut.
Node.	L. *nodus*, knot.
Notochord.	G. *noton*, back, + *chorde*, cord.
Oocyte.	G. *oon*, egg, + *kylos*, hollow body.
Oogenesis.	G. *oon*, egg, + *genesis*, birth.
Oogonium.	G. *oon*, egg, + *gonos*, offspring.
Organ.	L. *organum*, implement.

Orifice.	L. *orificium*, opening.
Os.	L. *os*, bone.
Osseous.	L. *os* (pl. *ossa*), bone.
Ossicle.	L. *ossiculum*, small bone.
Ossification.	L. *os*, bone, + *facere*, to make.
Osteone.	G. *osteon*, bone.
Osteocyte.	G. *osteon*, bone, + *kytos*, cell.
Osteology.	G. *osteon*, bone, + *logos*, treatise.
Osteolysis.	G. *osteon*, bone, + *lysis*, melting.
Ostium.	L. *ostium*, door.
Ovary.	L. *ovum*, egg.
Oviduct.	L. *ovum*, egg, + *ductus*, duct, tube.
Oviparous.	L. *ovum*, egg, + *parus*, giving birth by laying eggs.
Ovum (pl. **ova**).	L. *ovum*, egg.
Oxyntic.	G. *oxyntos*, making acid.
Oxytocin.	G. *oxys*, swift, + *tokos*, birth.
Pampiniform.	L. *pampineus*, full of or wrapped around with vine leaves, + *forma*, likeness.
Pancreas.	G. *pan*, all, + *kreas*, flesh.
Panniculus.	L. *panniculus*, small piece of cloth; hence covering of deeper tissues.
Paracentesis.	G. *para*, alongside of, + *kentesis*, puncture.
Paradidymis.	G. *para*, alongside of, + *didymos*, double, hence alongside the testes.
Paraganglion.	G. *para*, along, beside, + *ganglion*, knot.
Paramedian.	G. *para*, alongside of, + *mesos*, middle.
Parametrium.	G. *para*, alongside of, + *metra*, womb.
Paraplegia.	G. *para*, alongside of, + *plesso*, I strike.
Parasympathetic.	G. *para*, alongside of, + *sympathetikos*, sympathetic.
Paraxial.	G. *para*, alongside of, + L. *axis*, axle.

Parenchyma.	G. *parenchyma*, pouring out into the adjacent.
Paresis.	G. *paresis*, relaxation.
Parietal.	L. *paries*, wall.
Parieties.	L. *paries*, wall.
Paroophoron.	G. *para*, alongside of, + *oon*, egg, + *pherein*, to bear.
Parous.	L. *pario*, to bear.
Parthenogenesis.	G. *parthos*, virgin, + *genesis*, birth.
Penis.	L. *penis*, tail.
Pepsin.	G. *pepsis*, digestion.
Pepsinogen.	G. *pepsis*, digestion, + *gennao*, I produce.
Percussion.	L. *percussio*, striking.
Perichondrium.	G. *peri*, around, + *chondros*, cartilage.
Perimysium.	G. *peri*, around, + *mys*, muscle.
Perineum.	G. *perinaion*, perineum.
Periosteum.	G. *peri*, around, + *osteon*, bone.
Peristalsis.	G. *peristaltikos*, clasping and compressing.
Peritoneum.	G. *peri*, around, + *teinein*, to stretch.
Phallic.	G. *phallikos*, pertaining to the penis.
Phallus.	G. *phallos*, phallus; penis was a later meaning.
Phrenic.	G. *phren*, diaphragm.
Placenta.	L. *placenta*, a flat cake.
Plexus (pl. -us).	L. *plexus*, plaiting, braid.
Porta.	L. *porta* (pl. -ae), gate.
Portal.	L. *porta* (pl. -ae), gate.
Portio.	L. *portio*, part.
Pregnancy.	L. *prae*, before, + *gnascor*, to be born.
Prepuce.	L. *praeputium*, foreskin.
Processus.	L. *processus*, going forwards.
Proctodeum.	G. *proktos*, anus, + *hodaios*, pertaining to a way.
Promontory.	G. *promontorium*, mountain ridge.

53

Pronephros.	G. *pro*, before, + *nephros*, kidney.
Prostate.	L. *pro*, in front, + *stare*, to stand.
Psoas.	G. *psoa*, loin.
Pubes.	L. *pubes*, mature.
Pubis (pl. -es).	L. *pubes*, mature.
Pudendal.	L. *pudere*, to be ashamed.
Pyelogram.	G. *pyelos*, tub, trough, + *gramma*, mark.
Pyelograph.	G. *pyelos*, tub, trough, + *graphein*, to draw.
Pylorus.	G. *pylouros*, gate-keeper.
Ramify.	L. *ramus*, branch, + *facere*, to make.
Ramus.	L. *ramus*, branch.
Raphe.	G. *raphe*, seam.
Recess.	L. *recessus*, retreat.
Rectum, rectus.	L. *rectus*, straight.
Renal.	L. *renes*, kidneys.
Restiform.	L. *restis*, rope, + *forma*, shape, form.
Sac.	L. *saccus*, sack.
Sacculus.	L. *sacculus*, a little bag.
Salpinx.	G. *salpinx*, trumpet.
Sarcolemma.	G. *sarx*, flesh, + *lemma*, husk, skin.
Scoliosis.	G. *skoliosis*, curvature.
Scrotum.	L. *scrotum*, skin.
Semen.	L. *semen*, seed.
Semilunar.	L. *semi*, half, + *luna*, moon.
Seminiferous.	L. *semen*, seed, + *ferre*, to bear.
Septum.	L. *saeptum*, fence.
Serum.	L. *serum*, whey.
Sigmoid.	G. *sigma*, Greek letters, + *eidos*, shape, likeness.
Sinister.	L. *sinister*, left side or unlucky.
Sinus (pl. -us).	L. *sinus*, curve, cavity, bosom.

Sinusoid.	L. *sinus*, curve, cavity, + *eidos*, shape, likeness.
Situs inversus viscerum.	L. *situs*, site, position, + *inversus*, inverted, + *viscerum*, of the viscera.
Smegma.	G. *smegma*, soap.
Somatic.	G. *soma* (pl. *somata*), body.
Somatopleure.	G. *soma*, body, + *pleura*, side.
Somite.	G. *soma*, body, + suffix *-ite*, indicating origin.
Sperm.	G. *sperma*, seed.
Spermatocyte.	G. *sperma*, seed, + *kytos*, cell.
Spermatogenesis.	G. *sperma*, seed, + *genesis*, origin.
Spermatogonium.	G. *sperma*, seed, + *gone*, generation.
Spermatozoon (pl. -a).	G. *sperma*, seed, + *zoon*, animal.
Sphincter.	G. *sphingein*, to bind tight.
Splanchnic.	G. *splanchna*, viscera.
Splanchnology.	G. *splanknon*, viscus, + *logos*, treatise.
Spleen.	L. *splen*, spleen.
Spongiosum.	G. *spongia*, sponge.
Spongioblast.	G. *spongia*, sponge, + *blastos*, germ, bud.
Stenosis.	G. *stenosis*, narrowing.
Stomach.	G. *stomachos*, gullet, oesophagus.
Stomodeum, Stomatodeum.	G. *stoma*, mouth, + *odaios*, pertaining to a way.
Subcostal.	L. *sub*, under, + *costa*, rib.
Supracostal.	L. *supra*, above, + *costa*, rib.
Sympathetic.	G. *syn*, together, + *pathein*, to suffer.
Symphysis.	G. *syn*, together, + *physis*, growth.
Syndrome.	G. *syndrome*, occurrence.
Taenia, Tenia.	L. *taenia*, band, ribbon.
Terminalis.	L. *terminare*, to limit.
Testicle.	L. *testiculus*, testis.

Testis.	L. *testis*, testicle; a witness.
Theca.	L. *theca*, envelope, sheath.
Tract.	L. *tractus*, wool drawn out for spinning.
Trigone.	L. *trigonum*, triangle.
Truncus.	L. *truncus*, trunk of tree.
Tuberculum.	L. *tuberculum*, a small hump.
Tunica.	L. *tunica*, undergarment.
Umbilicus.	L. *umbilicus*, navel.
Urachus.	G. *ouron*, urine, + *cheo*, I pour.
Urea.	G. *ouron*, urine.
Ureter.	G. *ouron*, urine, + *tereo*, to preserve.
Urethra.	G. *ouethra*, word invented by Hippocrates (*c.* 460 B.C.).
Urine.	G. *ouron*, urine.
Urogenital.	G. *ouron*, urine, + L. *genitalis*, genital.
Urostyle.	G. *oura*, tail, + *stylos*, pillar.
Uterus.	L. *uterus*, womb.
Vagina.	L. *vagina*, sheath.
Valve.	L. *valva*, leaf of door.
Valvula.	L. *valvula*, a little fold, valve.
Vas.	L. *vas*, vessel.
Vein.	L. *vena*, vein.
Venter.	L. *venter*, belly.
Ventral.	L. *venter*, belly.
Vermiform.	L. *vermis*, worm, + *forma*, shape.
Vernix.	L. *vernatio*, shedding of snake skin.
Vertebra.	L. *vertebra*, joint.
Vertex.	L. *vertex*, whirl, whirlpool.
Vesica.	L. *vesica*, bladder.
Vesicle.	L. *vesicula*, a small bladder.
Vestibulum.	L. *vestibulum*, entrance court.

Vestige.   L. *vestigium*, trace, footprint.
Villus (pl. **villi**).   L. *villus*, hair.
Visceral.   L. *viscera*, entrails, bowels.
Viscus.   L. *viscus*, internal organ.
Vitelline.   L. *vitellus*, yolk of egg.
Vitellus.   L. *vitellus*, little calf (became transferred to yolk of egg by Celsus *c.* 10 A.D.).
Viviparous.   L. *vivus*, living, + *parere*, to beget.
Volvulus.   L. *volvulus*, twisted.
Vulva.   L. *volva*, cover.

Zona.   L. *zona*, belt, girdle.
Zonula.   L. dim. of *zona*, belt, girdle.
Zygapophysis.   G. *zygon*, yoke, + *apophysis*, process of a bone.

TERMS IMPORTANT IN THE HEAD AND NECK

Abducens.	L. *ab*, away, + *ducens*, leading.
Aberrant.	L. *ab*, away, + *errare*, to stray.
Accessory.	L. *accessorius*, supplementary.
Acinus.	L. *acinus*, grape.
Acoustic.	G. *akoustikos*, pertaining to hearing.
Acrania.	G. *a*, without, + *krania*, heads.
Adductor.	L. *ad*, to, + *ducere*, to lead.
Adenoid.	G. *aden*, acorn, + *eidos*, form, likeness.
Aditus.	L. *aditus*, opening.
Afferent.	L. *ad*, to, + *ferre*, to carry.
Agnathism.	G. *a*, without, + *gnathos*, jaw.
Ala.	L. *ala*, wing.
Ambiguus.	L. *amb*, both ways, + *agere*, to drive.
Ampulla.	L. *ampulla*, a flask or vessel swelling in the middle.
Amygdaloid.	G. *amydale*, almond, + *eidos*, form, likeness.
Anencephaly.	G. *an*, without, + *enkephalos*, brain.
Annulus.	L. *anulus* (*annulus*), a ring.
Ansa.	L. *ansa*, handle.
Antihelix.	G. *anti*, against, before, + *helix*, coil.
Antitragus.	G. *anti*, against, before, + *tragus*, goat, (perhaps the small tuft of hair growing here was likened to a goat's beard?).
Antrum.	G. *antron*, cave.
Aphasia.	G. *a*, without, + *phasis*, speech.

Aphonia.	G. *a*, without, + *phone*, voice.
Aqueduct.	L. *aqua*, water, + *ducere*, to lead.
Aqueous.	L. *aqua*, water.
Arachnoid.	G. *arachnes*, spider, + *eidos*, shape, likeness.
Arbor vitae.	L. *arbor*, tree, + *vitae*, of life.
Archipallium.	G. *archi*, principal, + *pallium*, cloak.
Arcuate.	L. *arcualis*, arch-shaped.
Arcus.	L. *arcus*, bow.
Arytenoid.	G. *arytana*, jug, + *eidos*, shape, likeness.
Asterion.	G. *aster*, star.
Ataxia.	G. *a*, without, + *taxis*, order.
Atresia.	G. *a*, without, + *tresis*, hole.
Atrium.	L. *atrium*, court, entrance hall.
Atrophy.	G. *a*, without, + *trophe*, nourishment.
Auditory.	L. *audire*, to hear.
Auricle.	L. dim., *auricula*, external ear.
Autonomic.	G. *autos*, self, + *nomos*, law.
Axial.	L. *axis*, axle of a wheel, the line about which any body turns.
Baroreceptor.	G. *baros*, weight (pressure receptor).
Basion.	G. *basis*, footing, base.
Basisphenoid.	G. *basis*, base, + *sphenoeides*, wedgeshaped.
Bicuspid.	L. *bis*, two, + *cuspis*, point.
Blepharal.	G. *blepharon*, eyelid.
Bolus.	G. *bolos*, mass.
Branchial.	L. b*ranchiae*, or G., *branchia*, gills.
Brachycephalic.	G. *brachys*, short, + *kephale*, head.
Bregma.	G. *brechein*, to moisten, (thought most humid part of infant skull).
Buccal.	L. *bucca*, cheek.

Bulbus.	L. *bulbus*, bulb, swollen root.
Bulla.	L. *bulla*, a bubble; hence spherical in shape.
Bursa.	L. *bursa*, a purse; hence purse-shaped object.
Callosum.	L. *callosus, -a, -um*, thick-skinned, also a beam or rafter.
Calvaria.	L. *calvus*, bald.
Canine.	L. *canis*, dog.
Canthus.	G. *kanthos*, rim on a wheel.
Caput (pl. capita).	L. *caput*, head.
Carotid.	G. *karon*, deep sleep (pressure on artery produces stupor).
Caruncle.	L. dim. of *caro*, flesh, any fleshy eminence.
Cava.	L. *cavus, -a, -um*, hollow or cave.
Cavernosus.	L. *caverna*, a hollow or cave.
Cavum.	L. *cavum*, a hollow or cave.
Cementum.	L. *caementum*, rough stone.
Centrum.	L. *centrum*, centre.
Cephalic.	G. *kephale*, head.
Cerebellum.	L. dim. *cerebrum*, brain.
Cerebrum.	L. *cerebrum*, brain.
Cervical.	L. *cervix*, neck.
Chiasma.	G. *chiasma*, figure of X.
Choana (pl. -ae).	G. *choane*, funnel.
Choroid.	G. *chorion*, skin, + *eidos*, shape, likeness.
Ciliary.	L. *cilium* (pl. *cilia*), eyelash.
Circumflex.	L. *circum*, around, + *flexere*, to bend.
Circumvallate.	L. *circum*, around, + *vallum*, wall.
Cisterna.	L. *cisterna*, reservoir.
Claustrum.	L. *claustrum*, barrier.
Cochlea.	L. *cochlea* (G., *kochlias*), spiral, snail shell.
Collum.	L. *collum*, neck.

Comes.	L. *comes*, companion.
Comitans	L. *comitari*, to accompany.
(pl. **comitantes**).	
Commissure.	L. *commissura* (*cum* + *mittere*), connection.
Communicans.	L. *communicans*, communicating.
Concha.	L. *concha*, bivalve, oyster shell.
Conjunctiva.	L. *conj*, together + *jungo*, I join, hence connecting.
Constrictor.	L. *constringere*, to draw together.
Conus.	L. *conus*, cone.
Convolution.	L. *con*, together, + *volvo*, to roll.
Copula.	L. *copula*, tie (from *copulare*, to copulate).
Cornea.	L. *corneus*, horny.
Cornu.	L. *cornu*, horn.
Corona.	L. *corona*, crown.
Coronary.	L. *coronarius*, pertaining to a wreath or crown; hence, encircling.
Coronoid.	G. *korax*, crow, + *eidos*, form, likeness.
Corpus	L. body.
(pl. **corpora**).	
Corrugator.	L. *con*, together + *ruga*, wrinkle.
Cortex	L. *cortex*, bark, rind.
(adj. **cortical**).	
Corticofugal.	L. *cortex*, bark of a tree, + *fugere*, to flee.
Corticopetal.	L. *cortex*, bark of a tree, + *petere*, to seek.
Cranial.	G. *kranion*, skull.
Cranium	G. *kranion*, skull.
(adj. **cranial**).	
Cribriform.	L. *cribrum*, sieve, + *forma*, form.
Cricoid.	G. *krikos*, ring, + *eidos*, form, likeness.
Crista.	L. *crista*, crest.
Crista galli.	L. *crista*, crest + L. *gallus*, cock, hence cock's comb.

61

Cruciate.	L. *crux*, cross.
Crus (pl. **crura**).	L. *crus*, leg.
Culmen.	L. *culmen*, top.
Cupula.	L. *cupula*, a small cask or cup.
Cuspis.	L. *cuspis*, point.
Dacryon.	G. *dakryon*, tear.
Decussation.	L. *decussatio*, intersection of two lines, as in Roman X.
Dendrite.	G. *dendros*, tree.
Dens.	L. *dens*, tooth.
Dentine.	L. *dens*, tooth.
Dentition.	L. *dentitis*, cutting of teeth.
Detrusor.	L. *detrudere*, to push down.
Diastema (pl. -ata).	G. *diastema*, interval.
Diencephalon.	G. *dia*, through, + *enkephalos*, brain.
Diploë.	G. *diploë*, fold.
Diplopia.	G. *diploos*, double, + *opsis*, vision.
Disc.	L. *discus*, disc.
Diverticulum.	L. *divertere*, to turn aside.
Dolichocephalic.	G. *dolichos*, long + G. *kephale*, head.
Duct.	L. *ducere*, to lead or draw.
Ductule.	L. *ducere*, to lead, dim.
Dura.	L. *dura*, hard, strong.
Eminence.	L. *e*, out, + *minere*, to jut.
Emissary.	L. *e*, out, + *mittere*, to send.
Encephalocoele, **Encephalocele.**	G. *enkephalos*, brain, + *koilos*, hollow.
Encephalon.	G. *en*, within, + *kephalos*, head.
Endocranium.	G. *endon,* within, + *kranion*, skull.
Endolymph.	G. *endon*, within, + L. *lympha*, water.
Ependyma.	G. *epi*, upon, + *endyma*, garment.

Epicanthus.	G. *epi*, upon, + *kanthos*, rim.
Epicranium.	G. *epi*, upon, + *kranion*, skull.
Epiglottis.	G. *epi*, upon, + *glottis*, larynx.
Epithalamus.	G. *epi*, on, + *thalamos*, chamber.
Ethmoid.	G. *ethmos*, sieve, + *eidos*, form, likeness.
Exophthalmos.	G. *ex*, out, + *ophthalmos*, eye.
Facialis.	L. *facies*, face.
Facies.	L. *facies*, face.
Falciform.	L. *falx*, sickle, + *forma*, shape, likeness.
Falx.	L. *falx*, *falcis*, sickle.
Fascia (pl. -iae).	L. *fascia*, band.
Fasciculus.	L. *fascis*, bundle.
Fastigius.	L. *fastigium*, slope.
Fauces.	L. *faux*, throat.
Fenestra.	L. *fenestra*, window.
Fissure.	L. *fissura* (*findo*), a cleft.
Flavum.	L. *flavus*, yellow.
Flocculus.	NL. dim. *floccus*, a tuft of wool.
Follicle.	L. *folliculus*, a small bag.
Fontanelle.	F. *fontanelle*, a small fountain.
Foramen.	L. *foramen*, hole.
Fornix.	L. *fornix*, arch or vault.
Fossa.	L. *fossa*, ditch, channel, something dug.
Fourchette.	F. *fourchette*, fork.
Fovea.	L. *fovea*, small pit.
Fraenum, Frenum.	L. *fraenum*, bridge.
Frenulum.	L. dim. *fraenum*, a little bridge.
Frontal.	L. *frons*, *frontis*, forehead, brow.
Fundus.	L. *fundus*, bottom.
Funicular.	L. *funis*, rope.
Fusiform.	L. *fusus*, spindle, + *forma*, shape, likeness.

Galea.	L. *galea*, leather helmet.
Gallus.	L. *gallus*, cock.
Genial.	G. *geneion*, chin.
Genioglossus.	G. *geneion*, chin, + *glossa*, tongue.
Geniohyoid.	G. *geneion*, chin, + *hyoeides*, U-shaped.
Glabella.	L. *glaber*, smooth.
Glia.	G. *gloia*, glue.
Globus.	L. *globus*, sphere.
Glossal.	G. *glossa*, tongue.
Glossopharyngeus.	G. *glossa*, tongue, + *pharynx*, throat.
Glottis.	G. *glottis*, mouth of the windpipe.
Gnathion.	G. *gnathos*, jaw.
Griseum.	L. *griseus*, bluish.
Gyrus (pl. gyri).	G. *gyros*, a turn.
Habenula.	L. *habena*, strap.
Hemiplegia.	G. *hemi*, half, + *plege*, stroke.
Hemisphere.	G. *hemi*, half, + *sphaira*, ball.
Hernia.	L. *hernia*, protrusion.
Hiatus.	L. *hiatus*, gap.
Hilum.	L. *hilum*, a small thing.
Hippocampus.	G. *hippos*, horse, + *kampos*, sea monster.
Hydrocephalus.	G. *hydor*, water, + *kephale*, head.
Hyoid.	G. *hyoeides*, U-shaped.
Hypaxial.	G. *hypo*, under, + L. *axis*, centre line, axis.
Hypoglossal.	G. *hypo*, under, + *glossa*, tongue.
Hypophysis.	G. *hypo*, under, + *physis*, growth.
Hypothalamus.	G. *hypo*, under, + *thalamos*, chamber, couch.
Icterus.	G. *ikteros*, jaundice.
Incisor.	L. *incisus* (*incidere*), cut.
Incisura.	L. *incidere*, to cut into.
Infundibulum.	L. *infundibulum*, a funnel.

Inion.	G. *inion*, back of head.
Insula.	L. *insula*, island.
Intima.	L. *intima*, innermost.
Intrinsic.	L. *intrinsecus*, inward.
Invagination.	L. *in*, in, + *vagina*, sheath.
Iris.	G. *iris*, rainbow.
Isthmus.	G. *isthmos*, narrow connection.
Jugal.	L. *jugum*, yoke.
Jugular.	L. *jugularis* (*jugulum*), pertaining to the neck.
Jugum.	L. *jugum*, yoke.
Labial.	L. *labialis* (*labia*), pertaining to lips.
Labium.	L. *labium*, lip.
Labrum.	L. *labrum*, rim.
Labyrinth.	G. *labyrinthos*, maze.
Lacerate.	L. *lacerare*, to tear.
Lacrimal.	L. *lacrima*, tear.
Lacuna.	L. *lacuna*, pond.
Lacus.	L. *lacus*, lake.
Lambdoid.	G. *lambda*, shaped like Greek letter L.
Lamella.	L. dim. of *lamina*, leaf.
Lamina (pl. -ae).	L. *lamina*, thin plate.
Larynx.	G. *larynx*, upper part of windpipe.
Leminiscus.	G. *leminiskos*, band.
Lemma.	G. *lemma*, skin.
Lens.	L. *lens*, lentil.
Leptomeninges.	G. *leptos*, thin, + *meninx*, membrane.
Limbus.	L. *limbus*, border.
Limen.	L. *limen*, threshold.
Ling-.	L. *lingua*, tongue.
Lingual.	L. *lingua*, tongue.
Lingula.	L. *lingula*, small tongue.

Lobus.	G. *lobos*, lobe.
Locus.	L. *locus*, place.
Lucidum.	L. *lucidus* (*lux*), full of light, clear.
Lumen.	L. *lumen*, light, opening.
Lunate.	L. *luna*, moon.
Macula.	L. *macula*, spot, stain.
Malar.	L. *mala*, cheek bone.
Malleus.	L. *malleus*, hammer.
Mandible.	L. *mando*, I chew.
Marginal.	L. *marginalis* (*margo*), bordering.
Masseter.	G. *masseter*, chewer.
Mastication.	L. *masticatio*, chewing.
Mastoid.	G. *mastos*, breast, + *eidos*, form, likeness.
Mater.	A. *mater*, wrap or nutrient.
Matrix.	L. *matrix* (*mater*), womb, groundwork, mold.
Maxilla.	L. *maxilla*, jaw, now bone of upper jaw.
Meatus (pl. **-us**).	L. *meatus*, passage.
Medulla.	L. *medulla*, marrow, pith.
Meninges.	G. *meninx*, membrane.
Meningocoele,	G. *meninx*, membrane, + *koilos*, hollow.
Meningocele.	
Meninx	G. *meninx*, membrane.
(pl. **meninges**).	
Mental.	L. *mentum*, chin.
Mesencephalon.	G. *mesos*, middle, + *en*, in, + *kephale*, head.
Metathalamus.	G. *meta*, after, + *thalamus*, chamber.
Metencephalon.	G. *meta*, after, + *enkephalos*, brain.
Metopic.	G. *metopon*, forehead.
Microcephaly.	G. *mikros*, small, + *kephale*, head.
Microdont.	G. *mikros*, small, + *odous*, tooth.
Microglia.	G. *mikros*, small, + *gloia*, glue.
Microsmatic.	G. *mikros*, small, + *osmaomai*, to smell.

Modiolus.	L. *modiolus*, hub.
Molar.	L. *molaris* (*mola*), pertaining to a millstone.
Myelencephalon.	G. *myelos*, marrow, + *enkephalos*, brain.
Mylohyoid.	G. *myle*, mill, + *hyoeides*, U-shaped.
Naris (pl. -es).	L. *naris*, nostril.
Nasion.	L. *nasus*, nose.
Neopallium.	G. *neos*, new, + *pallium*, cloak.
Nerve.	L. *nervus*, G. *neuron*, cordlike structure, nerve, tendon.
Neural.	G. *neuron*, nerve.
Neuralgia.	G. *neuron*, nerve, + *algos*, pain.
Neuraxon.	G. *neuron*, nerve, + *axon*, axis.
Neurilemma.	G. *neuron*, nerve, + *lemma*, husk, sheath.
Neuroblast.	G. *neuron*, nerve, + *blastos*, bud.
Neuroectomy.	G. *neuron*, nerve, + *ektome*, excision.
Neuroglia.	G. *neuron*, nerve, + *gloia*, glue.
Neurology.	G. *neuron*, nerve, + *logos*, treatise.
Neuron.	G. *neuron*, cordlike structure, sinew, tendon; equivalent of L. *nervus,* hence, nerve.
Nigra.	L. *niger*, black.
Nictitating.	L. *nictare*, to wink.
Node.	L. *nodus*, knot.
Notochord.	G. *noton*, back, + *chorde*, cord.
Nucha.	L. *nucha*, nape of neck.
Nuchal.	L. *nucha*, nape of neck.
Nystagmus.	G. *nystagma*, short sleep (during which rhythmic head movements occur), modern use, rhythmic eye movements.
Obelion.	G. *obelos*, pointed pillar.
Occipital.	L. *occipitium*, back of head.
Occlusion.	L. *ob*, before, + *claudo*, I close.

Oculomotor.	L. *oculus*, eye, + *motor*, mover.
Odontoblast.	G. *odous*, tooth, + *blastos*, bud.
Odontoid.	G. *odous*, tooth, + *eidos*, shape, likeness.
Oligodendroglia.	G. *oligos*, few, + *dendron*, tree, + *gloia*, glue.
Omohyoid.	G. *omos*, shoulder, + G. *hyoeides*, U-shaped; thus muscle passing from shoulder to U-shaped hyoid bone.
Opisthion.	G. *opisthios*, posterior.
Ophthalmic.	G. *ophthalmos*, eye.
Optic.	G. *opsis*, sight.
Oral.	L. *os*, *oris*, mouth; (compare L. *os*, bone, below).
Ora serrata.	L. *ora*, edge, + *serra*, saw.
Orbit.	L. *orbis*, anything circular.
Orifice.	L. *orificium*, opening.
Os.	L. *os*, bone.
Osseous.	L. *os* (pl. *ossa*), bone.
Ossicle.	L. *ossiculum*, small bone.
Ossification.	L. *os*, bone, + *facere*, to make.
Osteone.	G. *osteon*, bone.
Osteocyte.	G. *osteon*, bone, + *kytos*, cell.
Osteology.	G. *osteon*, bone, + *logos*, treatise.
Osteolysis.	G. *osteon*, bone, + *lysis*, melting.
Osteomalacia.	G. *osteon*, bone, + *malakia*, softness.
Ostium.	L. *ostium*, door.
Otic.	G. *otikos*, belonging to the ear.
Otolith.	G. *ous*, *otos*, ear, + *lithos*, stone.
Pachymeninges.	G. *pachys*, thick, + *meninx*, membrane.
Palate.	L. *palatum*, palate.
Pallidus.	L. *pallidus*, pale.
Pallium.	L. *pallium*, cloak.
Palpebra.	L. *palpebra*, eyelid.

Papilloedema, Papilledema.	L. *papilla*, nipple, + *oedema*, swelling.
Paraesthesia, Paresthesis.	G. *para*, alongside of, + *aisthesis*, sensation.
Paraflocculus.	G. *para*, alongside of, + L. *floccus*, flock of wool.
Paraganglion.	G. *para*, along, beside, + *ganglion*, knot.
Paralysis.	G. *para*, alongside of, + *lyein*, to loosen.
Paraplegia.	G. *para*, alongside of, + *plesso*, I strike.
Parasympathetic.	G. *para*, alongside of, + *sympathetikos*, sympathetic.
Parathyroid.	G. *para*, near, + *thyreos*, oblong shield, + *eidos*, form, likeness.
Paraxial.	G. *para*, alongside of, + L. *axis*, axle.
Paresis.	G. *paresis*, relaxation.
Parotid.	G. *para*, near, + *ous*, *otos*, ear.
Periodontal.	G. *peri*, around, + *odous*, tooth.
Petrosal.	L. *petrosus*, rocky.
Petrous.	L. *petrosus*, rocky.
Pharynx.	G. *pharynx*, throat.
Philtrum.	G. *philtron*, love potion, anything that awakens love.
Phonation.	G. *phone*, voice.
Pia.	L. *pius*, delicate.
Pilus.	L. *pilus*, hair.
Pineal.	L. *pinea*, pine cone.
Pinna (pl. -ae).	L. *penna*, *pinna*, feather; hence, wing.
Piriform.	L. *pirum*, pear, + *forma*, shape, likeness.
Pituitary.	L. *pituita*, slime, phlegm; at one time believed to secrete a mucous material from the brain into the nose.
Platycephaly.	G. *platys*, flat, + *kephale*, head.
Platysma.	G. *platysma*, plate.

Plexus (pl. **-us**).	L. *plexus*, plaiting, braid.
Plica.	L. *plicare*, to fold.
Pneumatic.	G. *pneumatikos*, pertaining to breath.
Pons.	L. *pons, pontis*, bridge.
Popliteus.	L. *poples*, ham.
Pore.	L. *porus*, passage.
Porta.	L. *porta* (pl. *-ae*), gate.
Portal.	L. *porta* (pl. *-ae*), gate.
Portio.	L. *portio*, part.
Porus.	G. *poros*, passage.
Premolar.	L. *pre*, in front, + *molaris*, molar.
Presbyopia.	G. *presbys*, old man, + *opsis*, sight.
Procerus.	L. *procerus*, tall, extended.
Processus.	L. *processus*, going forwards.
Prochordal.	G. *pro*, in front of, + *chorde*, cord.
Prognathism.	G. *pro*, in front of, + *gnathos*, jaw.
Promontory.	G. *promontorium*, mountain ridge.
Proptosis.	G. *pro*, before, + *ptosis*, falling.
Prosencephalon.	G. *pro*, before, + *enkephalos*, brain.
Prosthion.	G. *prosthen*, before.
Protuberance.	L. *protubero*, I swell.
Pterygoid.	G. *pteryx*, wing, + *eidos*, likeness, shape.
Ptosis.	G. *ptosis*, fall.
Pulvinar.	L. *pulvinar*, cushioned couch.
Punctum.	L. *punctum*, point.
Pupil.	L. *pupa*, girl.
Putamen.	L. *putamen*, shell, husk.
Pyramid.	G. *pyramis*, pyramid.
Pyriform.	L. *pirum*, pear, + *forma*, shape.
Quadrigeminus.	L. *quadrigeminus*, four-fold, four.

Recess.	L. *recessus*, retreat.
Reflect.	L. *reflecto*, to turn back.
Retina.	L. *rete*, net.
Rhinal.	G. *rhis*, nose.
Rhinencephalon.	G. *rhis*, nose, + *enkephalos*, brain.
Rhinion.	G. *rhinion*, nostril.
Rhombencephalon.	G. *rhombos*, rhomboid, + *enkephalos*, brain.
Rima.	L. *rima*, cleft.
Risorius.	L. *risus*, laughter.
Rostrum (pl. **-a**).	L. *rostrum*, beak.
Ruga (pl. **-ae**).	L. *ruga*, wrinkle.
Sac.	L. *saccus*, sack.
Sacculus.	L. *sacculus*, a little bag.
Salivary.	L. *saliva*, saliva.
Scala.	L. *scala*, staircase.
Scalene.	G. *skalenos*, uneven.
Scalp.	Teutonic. *skalp*, shell.
Sclera.	G. *skleros*, hard.
Sclerotic.	G. *skleros*, hard.
Sella turcica.	L. *sella*, saddle, + *turcica*, Turkish.
Semilunar.	L. *semi*, half, + *luna*, moon.
Septum.	L. *saeptum*, fence.
Sialogram.	G. *sialon*, saliva, + *gramma*, mark.
Sigmoid.	G. *sigma*, Greek letter S, + *eidos*, shape, likeness.
Sinister.	L. *sinister*, left side or unlucky.
Sinus (pl. **-us**).	L. *sinus*, curve, cavity, bosom.
Sphenoid.	G. *sphen*, wedge, + *eidos*, likeness, shape.
Stapes.	L. *stapes*, stirrup.
Strabismus.	G. *strabismos*, squinting.
Stratum (pl. **strata**).	L. *stratum*, layer.
Stria.	L. *stria*, furrow.

Striatum.	L. *striatus*, grooved, streaked.
Styloid.	G. *stylos*, pillar, + *eidos*, likeness.
Sublingual.	L. *sub*, under, + *lingua*, tongue.
Submandibular.	L. *sub*, under, + *mandibula*, jaw.
Substantia.	L. *substantia*, substance.
Sulcus.	L. *sulcus*, furrow.
Superciliary.	L. *super*, above, + *cilium*, eyelid.
Sustentaculum.	L. *sustentaculum*, support.
Sympathetic.	G. *syn*, together, + *pathein*, to suffer.
Taenia, Tenia.	L. *taenia*, band, ribbon.
Tapetum.	L. *tapete*, carpet, tapestry.
Tectum.	L. *tectum* (*tego*), roof.
Tegmen.	L. *tegmen*, covering.
Tegmentum.	L. *tegumentum*, covering.
Tela.	L. *tela*, web.
Telencephalon.	G. *telos*, end, + *enkephalos*, brain.
Tentorium.	L. *tentorium*, tent.
Teres.	L. *tero*, I grind, rub.
Terminalis.	L. *terminare*, to limit.
Thalamus.	G. *thalamos*, chamber.
Theca.	L. *theca*, envelope, sheath.
Tonsil.	L. *tonsilla*, mooring post.
Tragus.	G. *tragos*, goat, from small tuft of hair (goat's beard) in this region.
Tympanic.	L. *tympanum*, drum.
Uvea.	L. *uva*, grape.
Uvula.	L. *uva*, grape.
Vallate.	L. *vallum*, rampart, walled.
Vallecula.	L. dim. of *vallus*, fossa.
Valve.	L. *valva*, leaf of door.

Valvula.	L. *valvula*, a little fold, valve.
Velum.	L. *velum*, veil.
Ventricle.	L. *ventriculus*, little cavity, loculus.
Vermiform.	L. *vermis*, worm, + *forma*, shape.
Vermis.	L. *vermis*, worm.
Vertebra.	L. *vertebra*, joint.
Vertex.	L. *vertex*, whirl, whirlpool.
Vestibulum.	L. *vestibulum*, entrance court.
Vestige.	L. *vestigium*, trace, footprint.
Vitreus, vitreous.	L. *vitreus*, of glass; hence, transparent.
Vomer.	L. *vomer*, ploughshare.
Zygapophysis.	G. *zygon*, yoke, + *apophysis*, process of a bone.

TERMS SPECIFIC TO THE NERVOUS SYSTEM

Abducens.	L. *ab*, away, + *ducens*, leading.
Aberrant.	L. *ab*, away, + *errare*, to stray.
Accessory.	L. *accessorius*, supplementary.
Acoustic.	G. *akoustikos*, pertaining to hearing.
Aditus.	L. *aditus*, opening.
Afferent.	L. *ad*, to, + *ferre*, to carry.
Ala.	L. *ala*, wing.
Ambiguus.	L. *amb*, both ways, + *agere*, to drive.
Amygdaloid.	G. *amydale*, almond, + *eidos*, form, likeness.
Annulus.	L. *anulus* (*annulus*), a ring.
Ansa.	L. *ansa*, handle.
Antrum.	G. *antron*, cave.
Aphasia.	G. *a*, without, + *phasis*, speech.
Aphonia.	G. *a*, without, + *phone*, voice.
Apophysis.	G. *apo*, from. + *physis*, growth.
Aqueduct.	L. *aqua*, water, + *ducere*, to lead.
Aqueous.	L. *aqua*, water.
Arachnoid.	G. *arachnes*, spider, + *eidos*, shape, likeness.
Arbor vitae.	L. *arbor*, tree, + *vitae*, of life.
Archipallium.	G. *archi*, principal, + *pallium*, cloak.
Arcuate.	L. *arcualis*, arch-shaped.
Arcus.	L. *arcus*, bow.
Ataxia.	G. *a*, without, + *taxis*, order.
Auditory.	L. *audire*, to hear.

Autonomic.	G. *autos*, self, + *nomos*, law.
Axon.	G. *axon*, axis.
Baroreceptor.	G. *baros*, weight (pressure receptor).
Basilic.	A. *al-basilic*, inner vein.
Bulbus.	L. *bulbus*, bulb, swollen root.
Callosum.	L. *callosus, -a, -um*, thick-skinned, also a beam or rafter.
Calyx (pl. calices).	L. *calyx*, husk, cup-shaped protective covering.
Cauda, caudal.	L. *cauda*, tail.
Cava.	L. *cavus, -a, -um*, hollow or cave.
Cavernosus.	L. *caverna*, a hollow or cave.
Cavum.	L. *cavum*, a hollow or cave.
Celiac, Coeliac.	G. *koilia*, belly.
Cerebellum.	L. dim. *cerebrum*, brain.
Cerebrum.	L. *cerebrum*, brain.
Chiasma.	G. *chiasma*, figure of X.
Ciliary.	L. *cilium* (pl. *cilia*), eyelash.
Circumvallate.	L. *circum*, around, + *vallum*, wall.
Cisterna.	L. *cisterna*, reservoir.
Claustrum.	L. *claustrum*, barrier.
Cochlea.	L. *cochlea* (G., *kochlias*), spiral, snail shell.
Colliculus.	L. *colliculus*, little hill.
Commissure.	L. *commissura* (*cum* + *mittere*), connection.
Communicans.	L. *communicans*, communicating.
Concha.	L. *concha*, bivalve, oyster shell.
Convolution.	L. *con*, together, + *volvo*, to roll.
Copula.	L. *copula*, tie (from *copulare*, to copulate).
Corpus (pl. corpora).	L. body.

Cortex	L. *cortex*, bark, rind.
(adj. **cortical**).	
Corticofugal.	L. *cortex*, bark of a tree, + *fugere*, to flee.
Corticopetal.	L. *cortex*, bark of a tree, + *petere*, to seek.
Cotyledon.	G. *kotyledon*, cup-shaped hollow.
Culmen.	L. *culmen*, top.
Decussation.	L. *decussatio*, intersection of two lines, as in Roman X.
Dendrite.	G. *dendros*, tree.
Dermatome.	G. *derma*, skin, + *temnein*, to cut.
Diencephalon.	G. *dia*, through, + *enkephalos*, brain.
Diplopia.	G. *diploos*, double, + *opsis*, vision.
Diverticulum.	L. *divertere*, to turn aside.
Dura.	L. *dura*, hard, strong.
Dysarthria.	G. *dys*, hard, *arthron*, joint, hence disjointed speech.
Efferent.	L. *ex*, out, + *ferre*, to bear.
Encephalocoele,	G. *enkephalos*, brain, + *koilos*, hollow.
Encephalocele.	
Encephalon.	G. *en*, within, + *kephalos*, head.
Endolymph.	G. *endon*, within, + L. *lympha*, water.
Endoneurium.	G. *endon*, within, + *neuron*, nerve.
Ependyma.	G. *epi*, upon, + *endyma*, garment.
Epineurium.	G. *epi*, on, + *neuron*, nerve.
Epithalamus.	G. *epi*, on, + *thalamos*, chamber.
Falciform.	L. *falx*, sickle, + *forma*, shape, likeness.
Falx.	L. *falx*, *falcis*, sickle.
Fastigius.	L. *fastigium*, slope.
Fenestra.	L. *fenestra*, window.
Fibre, Fiber.	L. *fibra*, fibre, string, thread.

Fibril.	NL. *fibrilla*, a little thread.
Filament.	L. *filamentum*, thin fibre.
Filum.	L. *filum*, thread.
Fimbria.	L. *fimbriae*, threads, fringe.
Fissure.	L. *fissura* (*findo*), a cleft.
Flocculus.	NL. dim. *floccus*, a tuft of wool.
Follicle.	L. *folliculus*, a small bag.
Fornix.	L. *fornix*, arch or vault.
Fourchette.	F. *fourchette*, fork.
Fovea.	L. *fovea*, small pit.
Fundus.	L. *fundus*, bottom.
Funicular.	L. *funis*, rope.
Fusiform.	L. *fusus*, spindle, + *forma*, shape, likeness.
Ganglion.	G. *ganglion*, a swelling under the skin.
Genu.	L. *genu*, knee.
Glia.	G. *gloia*, glue.
Globus.	L. *globus*, sphere.
Glossopharyngeus.	G. *glossa*, tongue, + *pharynx*, throat.
Griseum.	L. *griseus*, bluish.
Gyrus (pl. **gyri**).	G. *gyros*, a turn.
Habenula.	L. *habena*, strap.
Hemiplegia.	G. *hemi*, half, + *plege*, stroke.
Hemisphere.	G. *hemi*, half, + *sphaira*, ball.
Hernia.	L. *hernia*, protrusion.
Hiatus.	L. *hiatus*, gap.
Hilum.	L. *hilum*, a small thing.
Hippocampus.	G. *hippos*, horse, + *kampos*, sea monster.
Hydrocephalus.	G. *hydor*, water, + *kephale*, head.
Hypaxial.	G. *hypo*, under, + L. *axis*, centre line, axis.
Hypoglossal.	G. *hypo*, under, + *glossa*, tongue.

Hypophysis.	G. *hypo*, under, + *physis*, growth.
Hypothalamus.	G. *hypo*, under, + *thalamos*, chamber, couch.
Impar.	L. *impar*, unequal.
Impression.	L. *in*, in, + *premere*, to press.
Incisura.	L. *incidere*, to cut into.
Indusium.	L. *indusium*, tunic.
Infundibulum.	L. *infundibulum*, a funnel.
Innervation.	L. *in*, in, + *nervus*, nerve.
Insula.	L. *insula*, island.
Intrinsic.	L. *intrinsecus*, inward.
Invagination.	L. *in*, in, + *vagina*, sheath.
Isthmus.	G. *isthmos*, narrow connection.
Jugular.	L. *jugularis* (*jugulum*), pertaining to the neck.
Jugum.	L. *jugum*, yoke.
Kinetic.	G. *kinesis*, movement.
Labyrinth.	G. *labyrinthos*, maze.
Lacuna.	L. *lacuna*, pond.
Lamina (pl. -ae).	L. *lamina*, thin plate.
Leminiscus.	G. *leminiskos*, band.
Leptomeninges.	G. *leptos*, thin, + *meninx*, membrane.
Limbus.	L. *limbus*, border.
Limen.	L. *limen*, threshold.
Lingula.	L. *lingula*, small tongue.
Lobus.	G. *lobos*, lobe.
Lucidum.	L. *lucidus* (*lux*), full of light, clear.
Lumen.	L. *lumen*, light, opening.
Lunate.	L. *luna*, moon.

Macula.	L. *macula*, spot, stain.
Mater.	A. *mater*, wrap or nutrient.
Matrix.	L. *matrix* (*mater*), womb, groundwork, mold.
Meatus (pl. -us).	L. *meatus*, passage.
Medulla.	L. *medulla*, marrow, pith.
Meninges.	G. *meninx*, membrane.
Meningocoele,	G. *meninx*, membrane, + *koilos*, hollow.
Meingocele.	
Meninx	G. *meninx*, membrane.
(pl. meninges).	
Mesencephalon.	G. *mesos*, middle, + *en*, in, + *kephale*, head.
Metathalamus.	G. *meta*, after, + *thalamus*, chamber.
Metencephalon.	G. *meta*, after, + *enkephalos*, brain.
Microcephaly.	G. *mikros*, small, + *kephale*, head.
Microglia.	G. *mikros*, small, + *gloia*, glue.
Microsmatic.	G. *mikros*, small, + *osmaomai*, to smell.
Mydriasis.	G. *mydriasis*, dilation of pupil.
Myelencephalon.	G. *myelos*, marrow, + *enkephalos*, brain.
Myenteric.	G. *mys*, muscle, + *enteron*, gut.
Neopallium.	G. *neos*, new, + *pallium*, cloak.
Nerve.	L. *nervus*, G. *neuron*, cordlike structure, nerve, tendon.
Neural.	G. *neuron*, nerve.
Neuralgia.	G. *neuron*, nerve, + *algos*, pain.
Neuraxon.	G. *neuron*, nerve, + *axon*, axis.
Neurenteric.	G. *neuron*, nerve, + *enteron*, gut.
Neurilemma.	G. *neuron*, nerve, + *lemma*, husk, sheath.
Neuroblast.	G. *neuron*, nerve, + *blastos*, bud.
Neuroectomy.	G. *neuron*, nerve, + *ektome*, excision.
Neuroglia.	G. *neuron*, nerve, + *gloia*, glue.
Neurology.	G. *neuron*, nerve, + *logos*, treatise.

Neuron.	G. *neuron*, cordlike structure, sinew, tendon; equivalent of L. *nervus*, hence, nerve.
Neuropil.	G. *neuron*, nerve, + *pilos*, felt.
Neuropore.	G. *neuron*, nerve, + *poros*, hole.
Nigra.	L. *niger*, black.
Node.	L. *nodus*, knot.
Nystagmus.	G. *nystagma*, short sleep (during which rhythmic head movements occur), modern use, rhythmic eye movements.
Oculomotor.	L. *oculus*, eye, + *motor*, mover.
Olfactory.	L. *olere*, to smell, + *facere*, to make.
Oligodendroglia.	G. *oligos*, few, + *dendron*, tree, + *gloia*, glue.
Ophthalmic.	G. *ophthalmos*, eye.
Optic.	G. *opsis*, sight.
Ostium.	L. *ostium*, door.
Otic.	G. *otikos*, belonging to the ear.
Otolith.	G. *ous, otos*, ear, + *lithos*, stone.
Pachymeninges.	G. *pachys*, thick, + *meninx*, membrane.
Pallidus.	L. *pallidus*, pale.
Pallium.	L. *pallium*, cloak.
Papilloedema, Papilledema.	L. *papilla*, nipple, + *oedema*, swelling.
Paraesthesia, Paresthesis.	G. *para*, alongside of, + *aisthesis*, sensation.
Paraflocculus.	G. *para*, alongside of, + L. *floccus*, flock of wool.
Paraganglion.	G. *para*, along, beside, + *ganglion*, knot.
Paralysis.	G. *para*, alongside of, + *lyein*, to loosen.
Paraplegia.	G. *para*, alongside of, + *plesso*, I strike.
Parasympathetic.	G. *para*, alongside of, + *sympathetikos*, sympathetic.

80

Paraxial.	G. *para*, alongside of, + L. *axis*, axle.
Parenchyma.	G. *parenchyma*, pouring out into the adjacent.
Paresis.	G. *paresis*, relaxation.
Parietal.	L. *paries*, wall.
Parieties.	L. *paries*, wall.
Pedicle.	L. *pes*, foot.
Pedunculus.	L. *pediculus*, a little foot.
Pellucida.	L. *per*, through, + *lucere*, to shine.
Perilymph.	G. *peri*, around, + L. *lympha*, fluid.
Perineum.	G. *perinaion*, perineum.
Petrosal.	L. *petrosus*, rocky.
Petrous.	L. *petrosus*, rocky.
Phonation.	G. *phone*, voice.
Photoreceptor.	G. *phos*, light, + *recipio*, I receive.
Phrenic.	G. *phren*, diaphragm.
Pia.	L. *pius*, delicate.
Piriform.	L. *pirum*, pear, + *forma*, shape, likeness.
Pituitary.	L. *pituita*, slime, phlegm; at one time believed to secrete a mucous material from the brain into the nose.
Placode.	G. *plax*, anything flat.
Plexus (pl. -us).	L. *plexus*, plaiting, braid.
Plica.	L. *plicare*, to fold.
Pons.	L. *pons*, *pontis*, bridge.
Presbyopia.	G. *presbys*, old man, + *opsis*, sight.
Processus.	L. *processus*, going forwards.
Prochordal.	G. *pro*, in front of, + *chorde*, cord.
Profundus.	L. *profundus*, deep.
Proprioceptor.	L. *proprius*, special, + *capere*, to take.
Proptosis.	G. *pro*, before, + *ptosis*, falling.
Prosencephalon.	G. *pro*, before, + *enkephalos*, brain.
Pulvinar.	L. *pulvinar*, cushioned couch.
Punctum.	L. *punctum*, point.

Pupil.	L. *pupa*, girl.
Putamen.	L. *putamen*, shell, husk.
Pyramid.	G. *pyramis*, pyramid.
Pyriform.	L. *pirum*, pear, + *forma*, shape.
Quadrigeminus.	L. *quadrigeminus*, four-fold, four.
Radicle.	L. *radix*, root.
Radix.	L. *radix*, root.
Ramify.	L. *ramus*, branch, + *facere*, to make.
Ramus.	L. *ramus*, branch.
Raphe.	G. *raphe*, seam.
Receptor.	L. *recipere*, to take back, receive.
Recess.	L. *recessus*, retreat.
Restiform.	L. *restis*, rope, + *forma*, shape, form.
Reticulum.	L. *reticulum*, little net.
Retina.	L. *rete*, net.
Rhinencephalon.	G. *rhis*, nose, + *enkephalos*, brain.
Rhombencephalon.	G. *rhombos*, rhomboid, + *enkephalos*, brain.
Rostrum (pl. -a).	L. *rostrum*, beak.
Sac.	L. *saccus*, sack.
Sacculus.	L. *sacculus*, a little bag.
Sarcolemma.	G. *sarx*, flesh, + *lemma*, husk, skin.
Scala.	L. *scala*, staircase.
Sciatic.	G. *ischion*, hip joint.
Sclera.	G. *skleros*, hard.
Sclerotic.	G. *skleros*, hard.
Septum.	L. *saeptum*, fence.
Sinister.	L. *sinister*, left side or unlucky.
Sinus (pl. -us).	L. *sinus*, curve, cavity, bosom.
Sinusoid.	L. *sinus*, curve, cavity, + *eidos*, shape, likeness.

Somatic.	G. *soma* (pl. *somata*), body.
Somite.	G. *soma*, body, + suffix *-ite*, indicating origin.
Spina.	L. *spina*, thorn.
Strabismus.	G. *strabismos*, squinting.
Stratum (pl. **strata**).	L. *stratum*, layer.
Stria.	L. *stria*, furrow.
Striatum.	L. *striatus*, grooved, streaked.
Sublingual.	L. *sub*, under, + *lingua*, tongue.
Submandibular.	L. *sub*, under, + *mandibula*, jaw.
Substantia.	L. *substantia*, substance.
Sulcus.	L. *sulcus*, furrow.
Superciliary.	L. *super*, above, + *cilium*, eyelid.
Sympathetic.	G. *syn*, together, + *pathein*, to suffer.
Synapse.	G. *syn*, together, + *haptein*, to fasten.
Tabular.	L. *tabula*, board, table.
Tapetum.	L. *tapete*, carpet, tapestry.
Tectum.	L. *tectum* (*tego*), roof.
Tegmen.	L. *tegmen*, covering.
Tegmentum.	L. *tegumentum*, covering.
Tela.	L. *tela*, web.
Telencephalon.	G. *telos*, end, + *enkephalos*, brain.
Tentorium.	L. *tentorium*, tent.
Terminalis.	L. *terminare*, to limit.
Thalamus.	G. *thalamos*, chamber.
Theca.	L. *theca*, envelope, sheath.
Tract.	L. *tractus*, wool drawn out for spinning.
Trigeminus.	L. *trigeminus*, born three together.
Tunica.	L. *tunica*, undergarment.
Unciform.	L. *uncus*, hook, + *forma*, shape.
Uncinate.	L. *uncinatus*, furnished with a hook.

Uncus.	L. *uncus*, hook.
Uvea.	L. *uva*, grape.
Vagus.	L. *vagus*, wandering.
Velum.	L. *velum*, veil.
Ventricle.	L. *ventriculus*, little cavity, loculus.
Vermis.	L. *vermis*, worm.
Vestibulum.	L. *vestibulum*, entrance court.
Zona.	L. *zona*, belt, girdle.
Zonula.	L. dim. of *zona*, belt, girdle.

ALPHABETICAL LIST OF TERMS

A

Abdomen.	L. from *abdere* (?), to hide.
Abducens.	L. *ab*, away, + *ducens*, leading.
Abductor.	L. *ab*, away, + *ducere*, to lead.
Aberrant.	L. *ab*, away, + *errare*, to stray.
Abortion.	L. *abortio*, to abort.
Absorption.	L. *absorptio*, to swallow.
Accessory.	L. *accessorius*, supplementary.
Acetabulum.	L. *acetabulum*, vinegar cup, shape of hip joint.
Acinus.	L. *acinus*, grape.
Acoustic.	G. *akoustikos*, pertaining to hearing.
Acrania.	G. *a*, without, + *krania*, heads.
Acromegaly.	G. *akros*, extremity, + *megale*, great.
Acromion.	G. *akron*, height, extremity, + *omos*, shoulder.
Adductor.	L. *ad*, to, + *ducere*, to lead.
Adenoid.	G. *aden*, acorn, + *eidos*, form, likeness.
Adhesion.	L. *adhaereo*, to stick together.
Adipose.	L. *adeps*, fat.
Aditus.	L. *aditus*, opening.
Adnexa.	L. *ad*, to, + *nexus*, bound.
Adolescence.	L. *adolesco*, to grow up.
Adrenal.	L. *ad*, near, + *renes*, kidneys.

Adventia.	L. *ad*, to, + *venire*, to come.
Afferent.	L. *ad*, to, + *ferre*, to carry.
Agnathism.	G. *a*, without, + *gnathos*, jaw.
Agonist.	G. *agonistes*, rival.
Ala.	L. *ala*, wing.
Allantois.	G. *allas*, sausage, + *eidos*, form, appearance.
Alveolus.	L. *alveolus*, little cavity.
Amastia.	G. *a*, without, + *mastos*, breast.
Ambiguus.	L. *amb*, both ways, + *agere*, to drive.
Amnion.	G. *amnion*, fetal membrane.
Amphiarthrosis.	G. *amphi*, on both sides, + *arthrosis*, joint.
Ampulla.	L. *ampulla*, a flask or vessel swelling in the middle.
Amygdaloid.	G. *amydale*, almond, + *eidos*, form, likeness.
Analogy.	G. *ana*, according to, + *logos*, treatise.
Anastomosis.	G. *anastomoein*, to bring to a mouth, cause to communicate.
Anatomy.	G. *ana*, apart, + *tennein*, to cut.
Anconeus.	G. *agkon*, elbow.
Androgen.	G. *andros*, man, + *gennan*, to produce.
Android.	G. *andros*, man, + *eidos*, form, likeness.
Anencephaly.	G. *an*, without, + *enkephalos*, brain.
Angiology.	G. *angeion*, vessel, + *logos*, treatise.
Angle.	L. *angulus*, angle.
Ankle.	L. *angulus*, angle.
Ankylosis.	G. *ankylosis*, stiffening of the joint.
Anlage.	Ger. *an*, on, + *legen*, to lay (a laying on — *primordium*, precursor).
Annulus.	L. *anulus* (*annulus*), a ring.
Anomaly.	G. *an*, without, + *nomos*, law.
Ansa.	L. *ansa*, handle.

Antagonist.	G. *anti*, against, *agonistes*, rival.
Antebrachium.	L. *ante*, before, + *brachium*, arm.
Anteflexion.	L. *ante*, before, + *flexere*, to bend.
Anterior.	L. *ante*, before.
Anteversion.	L. *ante*, before, + *versio*, turning.
Anthropoid.	G. *anthropos*, man, + *eidos*, form, likeness.
Antihelix.	G. *anti*, against, before, + *helix*, coil.
Antitragus.	G. *anti*, against, before, + *tragus*, goat, (perhaps the small tuft of hair growing here was likened to a goat's beard?).
Antrum.	G. *antron*, cave.
Anus.	L. *anus*, fundament.
Aorta.	G. *aeiro*, to raise.
Apertura.	L. *apertura*, opening.
Aphasia.	G. *a*, without, + *phasis*, speech.
Aphonia.	G. *a*, without, + *phone*, voice.
Apocrine.	G. *apo*, from, + *krinein*, to separate.
Aponeurosis.	G. *apo*, from, + *neuron*, tendon (earlier: nerve, later: tendon).
Apophysis.	G. *apo*, from, + *physis*, growth.
Appendicular.	adjectival form of appendix.
Appendix.	L. *appendere*, to hang upon.
Aqueduct.	L. *aqua*, water, + *ducere*, to lead.
Aqueous.	L. *aqua*, water.
Arachnoid.	G. *arachnes*, spider, + *eidos*, shape, likeness.
Arachnodactyli.	as previous, spider fingers, Marfans syndrome.
Arbor vitae.	L. *arbor*, tree, + *vitae*, of life.
Archipallium.	G. *archi*, principal, + *pallium*, cloak.
Arcuate.	L. *arcualis*, arch-shaped.
Arcus.	L. *arcus*, bow.
Areola.	L. *areola* (dim. of *area*, open space).

Arrector.	L. *arrigere*, to raise, thus arrectores pilorum, small muscles in skin that erect the hairs.
Artefact.	L. *arte*, by art, + *factus*, made.
Artery.	G. *aer*, air, + *terein*, to keep (L. *arteria*, windpipe). ancient belief that blood vessels contained air.
Articulation.	L. *artus*, joint, *articulatus*, little joint, pl. *articulationes*, joints.
Arytenoid.	G. *arytana*, jug, + *eidos*, shape, likeness.
Aspera.	L. *asper*, rough.
Asterion.	G. *aster*, star.
Asthenic.	G. *a*, without, + *sthenos*, strength.
Atavistic.	L. *atavus*, grandfather.
Ataxia.	G. *a*, without, + *taxis*, order.
Atelectasis.	G. *ateles*, incomplete, + *ectasis*, expansion.
Atlas.	G. *atlao*, I sustain (Greek god who bears the world).
Atresia.	G. *a*, without, + *tresis*, hole.
Atrium.	L. *atrium*, court, entrance hall.
Atrophy.	G. *a*, without, + *trophe*, nourishment.
Auditory.	L. *audire*, to hear.
Auricle.	L. dim., *auricula*, external ear.
Autonomic.	G. *autos*, self, + *nomos*, law.
Axial.	L. *axis*, axle of a wheel, the line about which any body turns.
Axilla.	L. *axilla*, armpit.
Axis.	L. *axis*, axle of a wheel.
Axon.	G. *axon*, axis.
Azygos.	G. *a*, not, + *zygon*, yoke; hence, unpaired.

B

Baroreceptor.	G. *baros*, weight (pressure receptor).
Basal.	L. *basis*, footing, base.
Basilic.	A. *al-basilic*, inner vein.
Basion.	G. *basis*, footing, base.
Basisphenoid.	G. *basis*, base, + *sphenoeides*, wedgeshaped.
Biceps.	L. *bis*, two, + *caput*, head.
Bicornuate.	L. *bis*, two, + *cornua*, horns.
Bicuspid.	L. *bis*, two, + *cuspis*, point.
Bifid.	L. *bis*, two, + *findere*, to cleave.
Bifurcate.	L. *bis*, two, + *furca*, fork.
Bilateral.	L. *bi*, two, + *latus*, side.
Bile.	L. *bilis*, bile.
Bipennate.	L. *bis*, two, + *pinna*, feather.
Biventer.	L. *bis*, two, + *venter*, belly.
Blepharal.	G. *blepharon*, eyelid.
Bolus.	G. *bolos*, mass.
Brachial.	L. *brachialis*, belonging to the arm (upper arm).
Brachium (pl. -ia).	L. *brachium*, arm, (upper arm).
Branchial.	L. b*ranchiae*, or G. *branchia*, gills.
Brachycephalic.	G. *brachys*, short, + *kephale*, head.
Bregma.	G. *brechein*, to moisten, (thought most humid part of infant skull).
Brevis.	L. *brevis*, short.
Bronchiole.	G. *bronchiolus* (dim. of *bronchus*, *brechein*, to moisten).
Bronchus.	G. *bronchia*, end of windpipe.
Buccal.	L. *bucca*, cheek.
Bulbus.	L. *bulbus*, bulb, swollen root.

Bulla.	L. *bulla*, a bubble; hence spherical in shape.
Bursa.	L. *bursa*, a purse; hence purse-shaped object.

C

Calcaneus.	L. *calcaneus*, heel.
Calcar.	L. *calcar*, spur.
Calcification.	L. *calx*, lime, + *facere*, to make.
Callosum.	L. *callosus*, -*a*, -*um*, thick-skinned, also a beam or rafter.
Callous.	L. as above, newly formed bone at fracture site.
Calvaria.	L. *calvus*, bald.
Calyx	L. *calyx*, husk, cup-shaped protective
(pl. **calices**).	covering.
Canaliculus.	L. *canalis*, watepipe.
Cancellous.	L. *cancelli*, latticework.
Canine.	L. *canis*, dog.
Cannula.	L. dim. *canna*, reed.
Canthus.	G. *kanthos*, rim on a wheel.
Capillary.	L. *capillaris*, pertaining to the hair.
Capitulum.	L. dim. *caput*, small head.
Capsule.	L. dim. of *capsa*, box.
Caput	L. *caput*, head.
(pl. **capita**).	
Cardiac.	G. *kardiakos* (*kardia*), pertaining to the heart.
Cardinal.	L. *cardinalis*, pertaining to a door hinge, on which something important or fundamental hinges.
Carina.	L. *carina*, keel.
Carneae.	L. *carneus*, fleshy.

Carnivorous.	L. *carno*, flesh, + *vorare*, to devour.
Carnosus.	L. *carnosus*, fleshy.
Carotid.	G. *karon*, deep sleep (pressure on artery produces stupor).
Carpus.	G. *karpos*, wrist.
Cartilage.	L. *cartilago*, gristle, cartilage.
Caruncle.	L. dim. of *caro*, flesh, any fleshy eminence.
Catheter.	G. *kata*, down, + *heimi*, thrust.
Cauda, caudal.	L. *cauda*, tail.
Cava.	L. *cavus*, *-a*, *-um*, hollow or cave.
Cavernosus.	L. *caverna*, a hollow or cave.
Cavum.	L. *cavum*, a hollow or cave.
Cecum, Caecum.	L. *caecus*, *-a*, *-um*, blind.
Celiac, Coeliac.	G. *koilia*, belly.
Celom (e), Coelom.	G. *koiloma*, a hollow.
Cementum.	L. *caementum*, rough stone.
Centrum.	L. *centrum*, centre.
Cephalic.	G. *kephale*, head.
Cerebellum.	L. dim. *cerebrum*, brain.
Cerebrum.	L. *cerebrum*, brain.
Cervical.	L. *cervix*, neck.
Chiasma.	G. *chiasma*, figure of X.
Chirurgery.	G. *cheir*, hand, + *ergon*, work, (hence surgery).
Choana **(pl. -ae).**	G. *choane*, funnel.
Choledochus.	G. *chole*, bile, + *dochos* (*dechomai*), container.
Chorda.	G. *chorde*, string of gut, cord.
Chorion.	G. *chorion*, skin.
Choroid.	G. *chorion*, skin, + *eidos*, shape, likeness.
Chromatolysis.	G. *chroma*, colour, + *lysis*, dissolution.

91

Chromosome.	G. *chroma*, colour, + *soma*, body.
Chyle.	G. *chylos*, juice.
Chyme.	G. *chymos*, juice.
Ciliary.	L. *cilium* (pl. *cilia*), eyelash.
Cingulum.	L. *cingulum*, girdle.
Circumflex.	L. *circum*, around, + *flexere*, to bend.
Circumvallate.	L. *circum*, around, + *vallum*, wall.
Cisterna.	L. *cisterna*, reservoir.
Claustrum.	L. *claustrum*, barrier.
Clava.	L. *clava*, club.
Clavicle.	L. *clavicula*, a little key.
Climacteric.	G. *klimakter*, top of ladder (from *klimax*, ladder).
Cloaca.	L. *cloaca*, sewer, drain.
Coccyx.	G. *kokkyx*, a cuckoo, hence a structure shaped like a cuckoo's bill.
Cochlea.	L. *cochlea* (G. *kochlias*), spiral, snail shell.
Collagen.	G. *kolla*, glue, + L. *gen*, begetter of.
Collateral.	L. *con*, together, + *latus*, side.
Colliculus.	L. *colliculus*, little hill.
Colloid.	G. *kolla*, glue, + *eidos*, likeness, shape.
Collum.	L. *collum*, neck.
Colon.	G. *kolon*, great gut.
Comes.	L. *comes*, companion.
Comitans (pl. comitantes).	L. *comitari*, to accompany.
Commissure.	L. *commissura* (*cum* + *mittere*), connection.
Communicans.	L. *communicans*, communicating.
Concha.	L. *concha*, bivalve, oyster shell.
Condyle.	G. *kondylos*, knuckle.
Conjunctiva.	L. *conj*, together + *jungo*, I join, hence connecting.
Constrictor.	L. *constringere*, to draw together.

Conus.	L. *conus*, cone.
Convolution.	L. *con*, together, + *volvo*, to roll.
Copula.	L. *copula*, tie (from *copulare*, to copulate).
Coraco-, corono-.	G. *korax* or *corone*, crow; hence crowlike.
Coracoid.	G. *korax*, crow, + *eidos*, form, likeness.
Corium.	G. *chorion*, skin, leather.
Cornea.	L. *corneus*, horny.
Cornu.	L. *cornu*, horn.
Corona.	L. *corona*, crown.
Coronary.	L. *coronarius*, pertaining to a wreath or crown; hence, encircling.
Coronoid.	G. *korax*, crow, + *eidos*, form, likeness.
Corpus (pl. **corpora**).	L. body.
Corpuscle.	L. *corpusculum*, little body.
Corrugator.	L. *con*, together + *ruga*, wrinkle.
Cortex (adj. **cortical**).	L. *cortex*, bark, rind.
Corticofugal.	L. *cortex*, bark of a tree, + *fugere*, to flee.
Corticopetal.	L. *cortex*, bark of a tree, + *petere*, to seek.
Costa (adj. **costal**).	L. *costa*, rib.
Cotyledon.	G. *kotyledon*, cup-shaped hollow.
Coxa.	L. *coxa*, hip.
Cranial	G. *kranion*, skull.
Cranium. (adj. **cranial**).	G. *kranion*, skull.
Cremaster.	G. *cremaster*, suspender.
Cribriform.	L. *cribrum*, sieve, + *forma*, form.
Cricoid.	G. *krikos*, ring, + *eidos*, form, likeness.
Crista.	L. *crista*, crest.
Crista galli.	L. *crista,* crest + L. *gallus*, cock, hence cock's comb.

Cruciate.	L. *crux*, cross.
Crus (pl. crura).	L. *crus*, leg.
Cubitus.	L. *cubitus*, elbow.
Cuboid.	G. *kuboeides*, cube-shaped.
Culmen.	L. *culmen*, top.
Cuneiform.	L. *cuneus*, wedge, + *forma*, shape, likeness.
Cupula.	L. *cupula*, a small cask or cup.
Cuspis.	L. *cuspis*, point.
Cutaneous.	L. *cutis*, skin.
Cuticle.	L. *cutis*, skin.
Cutis.	L. *cutis*, skin.
Cyst (adj. **cystic**).	G. *kystis*, bag, bladder, pouch.
Cytology.	G. *kytos*, cell, + *logos*, treatise.
Cytoplasm.	G. *kytos*, cell, + *plasma*, plasma, anything molded.

D

Dacryon.	G. *dakryon*, tear.
Dactyl.	G. *daktylos*, finger.
Dartos.	G. *dartos*, skinned.
Decidua (adj. **Deciduous**).	L. *deciduus*, (*de* + *cado*), falling off.
Decussation.	L. *decussatio*, intersection of two lines, as in Roman X.
Defaecation, Defecation.	L. *defaecare*, to cleanse.
Deferens.	L. *de*, away, + *ferens*, carrying.
Deglutition.	L. *deglutire*, to swallow.
Deltoid.	G. *delta*, letter in Greek alphabet, triangular-shaped.
Dendrite.	G. *dendros*, tree.
Dens.	L. *dens*, tooth.

Dentine.	L. *dens*, tooth.
Dentition.	L. *dentitis*, cutting of teeth.
Depressor.	L. *de*, down, + *premere*, to press.
Dermal.	G. *derma*, skin.
Dermatoglyphics.	G. *derma*, skin, + *glyphein*, to carve.
Dermatome.	G. *derma*, skin, + *temnein*, to cut.
Dermis.	G. *derma*, skin.
Detritus.	L. *deterere*, to rub away.
Detrusor.	L. *detrudere*, to push down.
Diagnosis.	G. *dia*, through, + *gnosis*, knowledge.
Diaphragm.	G. *dia*, through, + *phragma*, wall.
Diaphysis.	G. *dia*, between, + *physis*, growth.
Diarthrosis.	G. *dia*, through, + *arthroun*, to fasten by a joint.
Diastema	G. *diastema*, interval.
(pl. -ata).	
Diastole.	G. *dia*, through, + *stellein*, to send.
Diencephalon.	G. *dia*, through, + *enkephalos*, brain.
Digastric.	G. *dis*, double, + *gaster*, belly.
Digestion.	L. *dis*, apart, + *gerere*, to carry.
Digit.	L. *digitus*, finger.
Diploë.	G. *diploë*, fold.
Diploid.	G. *diploos*, twofold.
Diplopia.	G. *diploos*, double, + *opsis*, vision.
Disc.	L. *discus*, disc.
Distal.	L. *distare*, to stand apart.
Diverticulum.	L. *divertere*, to turn aside.
Dizygotic.	G. *dis*, twice, + *zygoo*, join together.
Dolichocephalic.	G. *dolichos*, long + G. *kephale*, head.
Dorsal.	L. *dorsum*, back.
Duct.	L. *ducere*, to lead or draw.
Ductule.	L. *ducere*, to lead, dim.

Duodenum.	L. *duodeni*, twelve (meaning twelve fingerbreadths).
Dura.	L. *dura*, hard, strong.
Dysarthria.	G. *dys*, hard, *arthron*, joint, hence disjointed speech.

E

Ectoderm.	G. *ektos*, outside, + *derma*, skin.
Ectopia.	G. *ektopos*, distant.
Effector.	L. *efficere*, to bring to pass.
Efferent.	L. *ex*, out, + *ferre*, to bear.
Ejaculatory.	L. *e*, out, + *jacere*, to throw.
Elastin.	G. *elastikos*, impulsive.
Emboliformis.	G. *embolos*, wedge, + L. *forma*, shape.
Embolus.	G. *embolos*, wedge, plug, anything inserted.
Embryo.	G. *embryon*, fruit in the womb, from G. *bryein*, to grow.
Embryology.	G. *embryon*, fruit in the womb, + *logos*, treatise.
Eminence.	L. *e*, out, + *minere*, to jut.
Emissary.	L. *e*, out, + *mittere*, to send.
Enarthrosis.	G. *en*, in, + *arthron*, joint.
Encephalocoele, Encephalocele.	G. *enkephalos*, brain, + *koilos*, hollow.
Encephalon.	G. *en*, within, + *kephalos*, head.
Endocardium.	G. *endon*, within, + *kardia*, heart.
Endochondrial.	G. *endon*, within, + *chondros*, cartilage.
Endocranium.	G. *endon*, within, + *kranion*, skull.
Endocrine.	G. *endon*, within, + *krinein*, to separate.
Endoderm.	G. *endon*, within, + *derma*, skin.
Endolymph.	G. *endon*, within, + L. *lympha*, water.
Endometrium.	G. *endon*, within, + *metra*, womb.

Endomysium.	G. *endon*, within, + *mys*, muscle.
Endoneurium.	G. *endon*, within, + *neuron*, nerve.
Endosteum.	G. *endon*, within, + *osteon*, bone.
Endoskeleton.	G. *endon*, within, + *skeletos*, dried up.
Endothelium.	G. *endon*, within, + *thele*, nipple.
Enzyme.	G. *en*, in, + *zyme*, leaven.
Ependyma.	G. *epi*, upon, + *endyma*, garment.
Epicanthus.	G. *epi*, upon, + *kanthos*, rim.
Epicardium.	G. *epi*, on, + *kardia*, heart.
Epicondyle.	G. *epi*, on, + *kondylos*, knuckle.
Epicranium.	G. *epi*, upon, + *kranion*, skull.
Epidermis.	G. *epi*, on, + *derma*, skin.
Epididymis.	G. *epi*, on, + *didmoi*, testicles.
Epigastrium.	G. *epi*, upon, + *gaster*, belly.
Epiglottis.	G. *epi*, upon, + *glottis*, larynx.
Epimysium.	G. *epi*, upon, + *mys*, muscle.
Epineurium.	G. *epi*, on, + *neuron*, nerve.
Epiphysis.	G. *epi*, on, + *physis*, growth.
Epiploic.	G. *epiploon*, caul, omentum.
Epispadias.	G. *epi*, upon, + N.L. *spadias*, F. *spadon*, eunuch, F. *span*, to draw.
Epithalamus.	G. *epi*, on, + *thalamos*, chamber.
Epithelium.	G. *epi*, on, + *thele*, nipple.
Epoöphoron.	G. *epi*, upon, + *oön*, egg, + *phoros*, bearing.
Erector.	L. *erectus*, erect.
Erythrocyte.	G. *erythros*, red, + *kytos*, cell.
Ethmoid.	G. *ethmos*, sieve, + *eidos*, form, likeness.
Eugenics.	G. *en*, well, + *gennan*, to generate.
Eversion.	L. *e*, out, + *vertere*, to turn.
Exomphalos.	G. *ex*, out, + *omphalos*, navel.
Excretion.	L. (*excretus*) *ex*, out, + *cernere*, to sift.
Exocrine.	G. *ex*, out, + *krinein*, to separate.

Exophthalmos.	G. *ex*, out, + *ophthalmos*, eye.
Exoskeleton.	G. *ex*, out, + *skeleton*, skeleton.
Extension.	L. *extendo*, extend.
Extensor.	L. *ex*, out, + *tendere*, to stretch.
Extravasation.	L. *extra*, outside, + *vas*, vessel.
Extrinsic.	L. *extrinsecus*, on the outside.
Exudate.	L. *ex*, out, + *sudare*, to sweat.

F

Facet.	F. *facette*, face.
Facialis.	L. *facies*, face.
Facies.	L. *facies*, face.
Faeces, Feces.	L. *faex*, dregs.
Falciform.	L. *falx*, sickle, + *forma*, shape, likeness.
Falx.	L. *falx, falcis*, sickle.
Fascia (pl. -iae).	L. *fascia*, band.
Fasciculus.	L. *fascis*, bundle.
Fastigius.	L. *fastigium*, slope.
Fauces.	L. *faux*, throat.
Femur.	L. *femur*, thigh.
Fenestra.	L. *fenestra*, window.
Fetus, Foetus.	L. *fetus*, offspring.
Fibre, Fiber.	L. *fibra*, fibre, string, thread.
Fibril.	NL. *fibrilla*, a little thread.
Fibroblast.	L. *fibra*, fibre, + *blastos*, bud.
Fibrocartilage.	L. *fibra*, fibre, + *cartilago*, gristle.
Fibula.	L. *fibula*, pin, skewer, brooch.
Filament.	L. *filamentum*, thin fibre.
Filum.	L. *filum*, thread.
Fimbria.	L. *fimbriae*, threads, fringe.
Fissure.	L. *fissura* (*findo*), a cleft.

Fistula.	L. *fistula*, pipe.
Flaccid.	L. *flaccidus*, weak.
Flavum.	L. *flavus*, yellow.
Flexor.	L. *flexus*, bent.
Flocculus.	NL. dim. *floccus*, a tuft of wool.
Folia.	L. *folium*, leaf.
Follicle.	L. *folliculus*, a small bag.
Fontanelle.	F. *fontanelle*, a small fountain.
Foramen.	L. *foramen*, hole.
Forceps.	L. *forceps*, pincers.
Form.	L. *forma*, shape.
Fornix.	L. *fornix*, arch or vault.
Fossa.	L. *fossa*, ditch, channel, something dug.
Fourchette.	F. *fourchette*, fork.
Fovea.	L. *fovea*, small pit.
Fraenum, Frenum.	L. *fraenum*, bridge.
Frenulum.	L. dim. *fraenum*, a little bridge.
Frontal.	L. *frons*, *frontis*, forehead, brow.
Fulcrum.	L. *fulcrum*, post.
Fundiform.	L. *funda*, sling, + *forma*, shape, likeness.
Fundus.	L. *fundus*, bottom.
Funicular.	L. *funis*, rope.
Fusiform.	L. *fusus*, spindle, + *forma*, shape, likeness.

G

Galea.	L. *galea*, leather helmet.
Gallus.	L. *gallus*, cock.
Gametes.	G. *gametes*, spouse.
Ganglion.	G. *ganglion*, a swelling under the skin.
Gastrocnemius.	G. *gaster*, belly, + *kneme*, leg.
Gastrula.	NL. dim. from G. *gaster*, stomach.

Gemellus.	L. *gemellus*, twin.
Genial.	G. *geneion*, chin.
Geniculate.	L. *geniculatus*, with bent knee.
Genioglossus.	G. *geneion*, chin, + *glossa*, tongue.
Geniohyoid.	G. *geneion*, chin, + *hyoeides*, U-shaped.
Genital.	L. *genitalis* (*gigno*), pertaining to birth.
Genu.	L. *genu*, knee.
Germinal.	L. *germen*, bud, germ.
Germinative.	L. *germen*, bud, germ.
Gestation.	L. *gestare*, to bear.
Ginglymus.	G. *ginglymus*, hinge.
Glabella.	L. *glaber*, smooth.
Gland.	L. *glandula*, dim. *glans*, acorn, pellet.
Glans.	L. *glans*, acorn.
Glenoid.	G. *glene*, socket, + *eidos*, form, likeness.
Glia.	G. *gloia*, glue.
Globus.	L. *globus*, sphere.
Glomerulus.	L. dim. of *glomus*, a ball.
Glossal.	G. *glossa*, tongue.
Glossopharyngeus.	G. *glossa*, tongue, + *pharynx*, throat.
Glottis.	G. *glottis*, mouth of the windpipe.
Gluteus.	G. *gloutos*, rump.
Gnathion.	G. *gnathos*, jaw.
Gomphosis.	G. *gomphos*, wedge-shaped nail.
Gonad.	G. *gone*, seed.
Gonion.	G. *gonia*, angle.
Gracilis.	L. *gracilis*, thin.
Granulation.	L. dim. *granum*, grain.
Granulum.	L. *granulum*, small grain.
Gravid.	L. *gravida*, pregnant.
Griseum.	L. *griseus*, bluish.
Gubernaculum.	L. *gubernaculum*, helm.

Gynaecomastia,	G. *gyne*, woman, + *mastos*, breast.
Gynecomastia.	G. *gyros*, a turn.
Gyrus (pl. gyri).	

H

Habenula.	L. *habena*, strap.
Habitus.	L. *habitus*, condition of the body.
Haemal, hemal.	G. *haima*, blood.
Hallux.	L. *hallux*, big toe.
Hamatum.	L. *hamatus*, hook-shaped.
Hamulus.	L. dim. *hamus*, hook.
Haploid.	G. *haplos*, plain.
Hemiplegia.	G. *hemi*, half, + *plege*, stroke.
Hemisphere.	G. *hemi*, half, + *sphaira*, ball.
Hemoglobin,	G. *haima*, blood, + L. *globus*, sphere.
Haemoglobin.	
Hemopoietic,	G. *haima*, blood, + *poietikos*, creative.
Haemopoietic.	
Hepar.	G. *hepar*, liver.
Hepatic.	L. *hepar*, liver.
Hernia.	L. *hernia*, protrusion.
Hiatus.	L. *hiatus*, gap.
Hilum.	L. *hilum*, a small thing.
Hippocampus.	G. *hippos*, horse, + *kampos*, sea monster.
Histology.	G. *histos*, web, + *logos*, treatise.
Holocrine.	G. *holos*, whole, + *krinein*, to separate.
Homology.	G. *homos*, same, + *logos*, treatise.
Hormone.	G. *hormaein*, to excite.
Humerus.	L. *humerus*, shoulder.
Humor.	L. *humor*, moisture, fluid.
Hyaline.	G. *hyalos*, glass.
Hydrocephalus.	G. *hydor*, water, + *kephale*, head.

Hydrocoele,	G. *hydro*, water, + *koilos*, hollow.
Hydrocele.	
Hymen.	G. *hymen*, membrane.
Hyoid.	G. *hyoeides*, U-shaped.
Hypaxial.	G. *hypo*, under, + L. *axis*, centre line, axis.
Hypoglossal.	G. *hypo*, under, + *glossa*, tongue.
Hypophysis.	G. *hypo*, under, + *physis*, growth.
Hypospadias.	G. *hypo*, under, + N.L. *spadias*, F. *spadon*, eunuch, F. *span*, to draw.
Hypothalamus.	G. *hypo*, under, + *thalamos*, chamber, couch.

I

Icterus.	G. *ikteros*, jaundice.
Ileum.	G. *eilein*, to wind or turn.
Iliacus.	See *ilium*.
Ilium.	L. *ileum*, flank.
Ima.	L. *ima*, lowest.
Impar.	L. *impar*, unequal.
Implantation.	L. *in*, in, + *plantere*, to plant.
Impression.	L. *in*, in, + *premere*, to press.
Incisor.	L. *incisus* (*incidere*), cut.
Incisura.	L. *incidere*, to cut into.
Incus.	L. *incus*, anvil.
Index.	L. *index*, pointer.
Inductor.	L. *inducere*, to lead on, excite.
Indusium.	L. *indusium*, tunic.
Infraspinous.	L. *infra*, below, + *spina*, spine.
Infundibulum.	L. *infundibulum*, a funnel.
Inguinal.	L. *inguina*, groin.
Inion.	G. *inion*, back of head.
Innervation.	L. *in*, in, + *nervus*, nerve.

Innominate.	L. *in*, in, + *nomen*, name, unnamed bone, unnamed artery.
Insertion.	L. *in*, in, + *serere*, to plant.
Insula.	L. *insula*, island.
Integument.	L. *in*, over, + *tegere*, to cover.
Intercalated.	L. *inter*, between, + *calare*, to call, inserted or placed between.
Intercostal.	L. *inter*, between, + *costa*, rib.
Interdigitating.	L. *inter*, between, + *digitus*, digit, interlocked by finger-like processes.
Interstitial.	L. *inter*, between, + *sistere*, to set, thus placed in spaces or gaps.
Intervertebral.	L. *inter*, between, + *vertebra*, joint.
Intestine.	L. *intus*, within, or L. *intestina*, guts, entrails.
Intima.	L. *intima*, innermost.
Intrinsic.	L. *intrinsecus*, inward.
Introitus.	L. *intro*, within, + *ire*, to go.
Invagination.	L. *in*, in, + *vagina*, sheath.
Inversion.	L. *in*, in, + *vertere*, to turn.
Iris.	G. *iris*, rainbow.
Ischiofemoral.	G. *ischion*, hip, + L. *femur*, thigh.
Ischium (pl. -ia).	G. *ischion*, hip.
Isthmus.	G. *isthmos*, narrow connection.

J

Jejunum.	L. *jejunus*, empty (of food), that part of intestine that appears empty.
Jugal.	L. *jugum*, yoke.
Jugular.	L. *jugularis* (*jugulum*), pertaining to the neck.
Jugum.	L. *jugum*, yoke.
Juxtaposition.	L. *juxta*, near, + *positio*, place.

K

Karyocyte.	G. *karyon*, nucleus, nut, + *kytos*, cell.
Karyolysis.	G. *karyon*, nucleus, nut, + *lysis*, loosening.
Keratin.	G. *keras*, horn.
Kinetic.	G. *kinesis*, movement.
Kyphosis.	G. *kyphos*, bent.

L

Labial.	L. *labialis* (*labia*), pertaining to lips.
Labium.	L. *labium*, lip.
Labrum.	L. *labrum*, rim.
Labyrinth.	G. *labyrinthos*, maze.
Lacerate.	L. *lacerare*, to tear.
Lacrimal.	L. *lacrima*, tear.
Lactation.	L. *lactare*, to suckle.
Lacteals.	L. *lactare*, to suckle.
Lacuna.	L. *lacuna*, pond.
Lacus.	L. *lacus*, lake.
Lambdoid.	G. *lambda*, shaped like Greek letter L.
Lamella.	L. dim. of *lamina*, leaf.
Lamina (pl. -ae).	L. *lamina*, thin plate.
Lanugo.	L. *lanugo*, fine hair.
Laparotomy.	L. *lapara*, flank, + *tome*, section, cut.
Larynx.	G. *larynx*, upper part of windpipe.
Lateral.	L. *lateralis* (*latus*), pertaining to a side.
Latissimus.	L. superlative, *latus*, broad, therefore very broad.
Leminiscus.	G. *leminiskos*, band.
Lemma.	G. *lemma*, skin.

Lens.	L. *lens*, lentil.
Leptomeninges.	G. *leptos*, thin, + *meninx*, membrane.
Leucocyte,	G. *leukos*, white, + *kytos*, cell.
Leukocyte.	
Levator.	L. *levare*, to raise.
Lien.	L. *lien*, spleen.
Lieno-.	L. *lien*, spleen from original L. *splen* (the sp having been dropped).
Ligamentum.	L. *ligamentum*, a bandage.
Ligature.	L. *ligare*, to bind.
Limbus.	L. *limbus*, border.
Limen.	L. *limen*, threshold.
Ling-.	L. *lingua*, tongue.
Lingual.	L. *lingua*, tongue.
Lingula.	L. *lingula*, small tongue.
Lipid.	G. *lipos*, fat, + *eidos*, resemblance.
Liver.	AS. *lifer*, liver.
Lobus.	G. *lobos*, lobe.
Locus.	L. *locus*, place.
Longus.	L. *longus*, long.
Lordosis.	G. *lordoo*, I bend.
Lucidum.	L. *lucidus* (*lux*), full of light, clear.
Lumbar.	L. *lumbare*, apron for the loins.
Lumbrical.	L. *lumbricus*, earthworm.
Lumen.	L. *lumen*, light, opening.
Lunate.	L. *luna*, moon.
Lung.	AS. *lunge*, lung.
Luteum.	L. *luteus*, yellow.
Lymph.	L. *lympha*, spring water.
Lymphocyte.	L. *lympha*, spring water, + G. *kytos*, cell.

M

Macrophage.	G. *makros*, large, + *phagein*, to eat.
Macroscopic.	G. *makros*, large, + *skopeo*, I see.
Macula.	L. *macula*, spot, stain.
Magnus, -a, -um.	L. *magnus*, large.
Malar.	L. *mala*, cheek bone.
Malleus.	L. *malleus*, hammer.
Mammary.	L. *mamma*, breast.
Mammillary.	L. dim. *mammillaris* (*mamma*, -ae), little breast.
Mandible.	L. *mando*, I chew.
Manubrium.	L. *manubrium*, handle, hilt (as of a sword).
Manus.	L. *manus*, hand.
Marginal.	L. *marginalis* (*margo*), bordering.
Masseter.	G. *masseter*, chewer.
Mastication.	L. *masticatio*, chewing.
Mastoid.	G. *mastos*, breast, + *eidos*, form, likeness.
Mater.	A. *mater*, wrap or nutrient.
Matrix.	L. *matrix* (*mater*), womb, groundwork, mold.
Maxilla.	L. *maxilla*, jaw, now bone of upper jaw.
Meatus (pl. -us).	L. *meatus*, passage.
Medial.	L. *medialis* (*medius*), pertaining to the middle.
Median.	L. *medianus*, in the middle.
Mediastinum.	L. *mediastinus*, servant, drudge, but anatomical term mediastinum probably derived from *per medium tensum*, tight in the middle.
Medulla.	L. *medulla*, marrow, pith.
Megaloblast.	G. *megas*, big, + *blastos*, bud.
Meiosis.	G. *meion*, less.
Melanin.	G. *melas*, black.
Membrane.	L. *membrana*, skin.

Meninges.	G. *meninx*, membrane.
Meningocoele,	G. *meninx*, membrane, + *koilos*, hollow.
Meningocele.	
Meniscus.	L. *menis*, cresent, half-moon, dim. of *mene*, moon.
Meninx	G. *meninx*, membrane.
(pl. **meninges**).	
Menopause.	G. *men*, month, + *pausis*, cessation.
Menstruation.	G. *men*, month, L. *menstruus*, pertaining to a month.
Mental.	L. *mentum*, chin.
Merocrine.	G. *meros*, portion, + *krinein*, to separate.
Mesencephalon.	G. *mesos*, middle, + *en*, in, + *kephale*, head.
Mesenchyme.	G. *mesos*, middle, + *en*, in, + *chymos*, juice.
Mesentery.	G. *mesos*, midway between, + *enteron*, gut.
Mesocolon.	G. *mesos*, middle, + *kolon*, great gut.
Mesoderm.	G. *mesos*, middle, + *derma*, skin.
Mesogastrium.	G. *mesos*, middle, + *gaster*, stomach.
Mesonephros.	G. *mesos*, middle, + *nephros*, kidney.
Mesosalpinx.	G. *mesos*, middle, + *salpinx*, tube.
Mesothelium.	G. *mesos*, middle, + *thele*, nipple, hence middle lining layer.
Mesovarium.	G. *mesos*, middle, + L. *ovarium*, ovary.
Metabolic.	G. *metabole*, change.
Metacarpus.	G. *meta*, after, + L. *carpus*, wrist.
Metamorphosis.	G. *meta*, after + *morphe*, form, signifying change.
Metanephros.	G. *meta*, after, + *nephros*, kidney.
Metaphase.	G. *meta*, after, + *phasis*, appearance.
Metaphysis.	G. *meta*, after, + *physis*, growth.
Metaplasia.	G. *meta*, after, + *plasma*, formed, molded.
Metatarsus.	G. *meta*, after, + L. *tarsus*, ankle.
Metathalamus.	G. *meta*, after, + *thalamus*, chamber.

Metencephalon.	G. *meta*, after, + *enkephalos*, brain.
Metopic.	G. *metopon*, forehead.
Microcephaly.	G. *mikros*, small, + *kephale*, head.
Microdont.	G. *mikros*, small, + *odous*, tooth.
Microglia.	G. *mikros*, small, + *gloia*, glue.
Microscope.	G. *mikros*, small, + *skopeo*, I look.
Microsmatic.	G. *mikros*, small, + *osmaomai*, to smell.
Microsome.	G. *mikros*, small, + *soma*, body.
Microtome.	G. *mikros*, small, + *tome*, cutting.
Minimus.	L. *minimus*, least.
Mitochondria.	G. *mitos*, thread, + *chondrion*, grain.
Mitosis.	G. *mitos*, thread.
Mitral.	L. *mitra*, turban, but Hebrew priest headress the only head covering that the mitral valve resembles.
Modiolus.	L. *modiolus*, hub.
Molar.	L. *molaris* (*mola*), pertaining to a millstone.
Mono-.	G. *monos*, alone.
Monocyte.	G. *monos*, single, + *kytos*, cell.
Mons.	L. *mons*, mountain.
Morph-.	G. *morphe*, shape.
Morphology.	G. *morphos*, form, + *logos*, treatise.
Morula.	L. *morus*, mulberry.
Mucus.	L. *mucus*, G. *muxa*, snivel, slippery secretion.
Multifidus.	L. *multifidus*, much divided.
Multiparous.	L. *multus*, many, + *parire*, to give birth.
Muscle.	L. *musculus*, little mouse.
Mydriasis.	G. *mydriasis*, dilation of pupil.
Myelencephalon.	G. *myelos*, marrow, + *enkephalos*, brain.
Myelin.	G. *myelos*, marrow.
Myenteric.	G. *mys*, muscle, + *enteron*, gut.
Mylohyoid.	G. *myle*, mill, + *hyoeides*, U-shaped.
Myocardium.	G. *mys*, *myos*, muscle, + *kardia*, heart.

Myology.	G. *mys*, muscle, + *logos*, treatise.
Myotome.	G. *mys*, *myos*, muscle, + *tome*, a cutting.
Myxoedema,	G. *myxa*, mucus, + *oidema*, swelling.
Myxedema.	

N

N.A.	Nomina Anatomica.
Naris (pl. **-es**).	L. *naris*, nostril.
Nasion.	L. *nasus*, nose.
Nates.	L. *nates*, buttocks.
Navel.	AS. *nafe*, centre of hub of wheel.
Navicular.	L. *navicula*, small boat.
Necrosis.	G. *nekrosis*, a killing.
Necropsy.	G. *nekros*, corpse, + *opsis*, sight.
Neo-.	G. *neos*, new.
Neopallium.	G. *neos*, new, + *pallium*, cloak.
Nerve.	L. *nervus*, G. *neuron*, cordlike structure, nerve, tendon.
Neural.	G. *neuron*, nerve.
Neuralgia.	G. *neuron*, nerve, + *algos*, pain.
Neuraxon.	G. *neuron*, nerve, + *axon*, axis.
Neurenteric.	G. *neuron*, nerve, + *enteron*, gut.
Neurilemma.	G. *neuron*, nerve, + *lemma*, husk, sheath.
Neuroblast.	G. *neuron*, nerve, + *blastos*, bud.
Neuroectomy.	G. *neuron*, nerve, + *ektome*, excision.
Neuroglia.	G. *neuron*, nerve, + *gloia*, glue.
Neurology.	G. *neuron*, nerve, + *logos*, treatise.
Neuron.	G. *neuron*, cordlike structure, sinew, tendon; equivalent of L. *nervus*, hence, nerve.
Neuropil.	G. *neuron*, nerve, + *pilos*, felt.
Neuropore.	G. *neuron*, nerve, + *poros*, hole.
Nictitans.	L. *nictare*, to wink.

Nigra.	L. *niger*, black.
Nictitating.	L. *nictare*, to wink.
Node.	L. *nodus*, knot.
Norma.	L. *norma*, rule, square used by carpenters, hence standard viewpoint.
Normoblast.	L. *norma*, rule, + *blastos*, bud.
Notochord.	G. *noton*, back, + *chorde*, cord.
Nucha.	L. *nucha*, nape of neck.
Nuchal.	L. *nucha*, nape of neck.
Nucleolus.	L. *nucleus*, nut. New derivation "small nucleus."
Nucleus.	L. *nucleus*, nut.
Nystagmus.	G. *nystagma*, short sleep (during which rhythmic head movements occur), modern use, rhythmic eye movements.

O

Obelion.	G. *obelos*, pointed pillar.
Obliquus.	L. *obliquus*, slanting.
Obturator.	L. *obturo*, I stop up.
Occipital.	L. *occipitium*, back of head.
Occlusion.	L. *ob*, before, + *claudo*, I close.
Oculomotor.	L. *oculus*, eye, + *motor*, mover.
Odontoblast.	G. *odous*, tooth, + *blastos*, bud.
Odontoid.	G. *odous*, tooth, + *eidos*, shape, likeness.
Oedema, Edema.	G. *oidema*, swelling.
Oesophagus, Esophagus.	G. *oisein* (*phero*), to carry, + *phagein*, to eat.
Oestrus, Estrus.	G. *oistros*, stinging insect, stung into activity at time of heat.
Olecranon.	G. *olene*, elbow, + *kranion*, head.
Olfactory.	L. *olere*, to smell, + *facere*, to make.

Oligodendroglia.	G. *oligos*, few, + *dendron*, tree, + *gloia*, glue.
Omentum.	L. *omentum*, adipose membrane enclosing the bowels.
Omohyoid.	G. *omos*, shoulder, + G. *hyoeides*, U-shaped; thus muscle passing from shoulder to U-shaped hyoid bone.
Ontogeny.	G. *onta*, things that exist, + *gennan*, to beget.
Oocyte.	G. *oon*, egg, + *kylos*, hollow body.
Oogenesis.	G. *oon*, egg, + *genesis*, birth.
Oogonium.	G. *oon*, egg, + *gonos*, offspring.
Operculum.	L. *operculum*, corner or lid.
Opisthion.	G. *opisthios*, posterior.
Ophthalmic.	G. *ophthalmos*, eye.
Opposition.	L. *ob*, in the way of, + *positus*, placed.
Optic.	G. *opsis*, sight.
Oral.	L. *os*, *oris*, mouth; (compare L. *os*, bone, below).
Ora serrata.	L. *ora*, edge, + *serra*, saw.
Orbit.	L. *orbis*, anything circular.
Organ.	L. *organum*, implement.
Orifice.	L. *orificium*, opening.
Os.	L. *os*, bone.
Osseous.	L. *os* (pl. *ossa*), bone.
Ossicle.	L. *ossiculum*, small bone.
Ossification.	L. *os*, bone, + *facere*, to make.
Osteone.	G. *osteon*, bone.
Osteocyte.	G. *osteon*, bone, + *kytos*, cell.
Osteology.	G. *osteon*, bone, + *logos*, treatise.
Osteolysis.	G. *osteon*, bone, + *lysis*, melting.
Osteomalacia.	G. *osteon*, bone, + *malakia*, softness.
Ostium.	L. *ostium*, door.
Otic.	G. *otikos*, belonging to the ear.
Otolith.	G. *ous*, *otos*, ear, + *lithos*, stone.

Ovary.	L. *ovum*, egg.
Oviduct.	L. *ovum*, egg, + *ductus*, duct, tube.
Oviparous.	L. *ovum*, egg, + *parus*, giving birth by laying eggs.
Ovum (pl. ova).	L. *ovum*, egg.
Oxyntic.	G. *oxyntos*, making acid.
Oxytocin.	G. *oxys*, swift, + *tokos*, birth.

P

Pachymeninges.	G. *pachys*, thick, + *meninx*, membrane.
Palate.	L. *palatum*, palate.
Pallidus.	L. *pallidus*, pale.
Pallium.	L. *pallium*, cloak.
Palma.	L. *palma*, the (open) hand.
Palpebra.	L. *palpebra*, eyelid.
Pampiniform.	L. *pampineus*, full of or wrapped around with vine leaves, + *forma*, likeness.
Pancreas.	G. *pan*, all, + *kreas*, flesh.
Panniculus.	L. *panniculus*, small piece of cloth; hence covering of deeper tissues.
Papilla.	L. *papilla*, nipple of breast.
Papilloedema, Papilledema.	L. *papilla*, nipple, + *oedema*, swelling.
Paracentesis.	G. *para*, alongside of, + *kentesis*, puncture.
Paradidymis.	G. *para*, alongside of, + *didymos*, double, hence alongside the testes.
Paradox.	G. *para*, alongside of, + *doxa*, belief.
Paraesthesia, Paresthesis.	G. *para*, alongside of, + *aisthesis*, sensation.
Paraflocculus.	G. *para*, alongside of, + L. *floccus*, flock of wool.

Paraganglion.	G. *para*, along, beside, + *ganglion*, knot.
Paralysis.	G. *para*, alongside of, + *lyein*, to loosen.
Paramedian.	G. *para*, alongside of, + *mesos*, middle.
Parametrium.	G. *para*, alongside of, + *metra*, womb.
Paraplegia.	G. *para*, alongside of, + *plesso*, I strike.
Parasternal.	G. *para*, alongside of, + *sternon*, chest.
Parasympathetic.	G. *para*, alongside of, + *sympathetikos*, sympathetic.
Parathyroid.	G. *para*, near, + *thyreos*, oblong shield, + *eidos*, form, likeness.
Paraxial.	G. *para*, alongside of, + L. *axis*, axle.
Parenchyma.	G. *parenchyma*, pouring out into the adjacent.
Paresis.	G. *paresis*, relaxation.
Parietal.	L. *paries*, wall.
Parieties.	L. *paries*, wall.
Paroophoron.	G. *para*, alongside of, + *oon*, egg, + *pherein*, to bear.
Parotid.	G. *para*, near, + *ous*, *otos*, ear.
Parous.	L. *pario*, to bear.
Pars.	L. *pars*, part.
Parthenogenesis.	G. *parthos*, virgin, + *genesis*, birth.
Patella.	L. *patella*, small pan.
Pathology.	G. *pathos*, disease, + *logos*, treatise.
Patulous.	L. *patulus*, standing open.
Pecten.	L. *pecten*, comb.
Pectoral.	L. *pectoralis* (*pectus*), belonging to the breast.
Pedicle.	L. *pes*, foot.
Pedunculus.	L. *pediculus*, a little foot.
Pellucida.	L. *per*, through, + *lucere*, to shine.
Pelvic.	L. *pelvis*, basin.
Penis.	L. *penis*, tail.

Penniform.	L. *penna*, feather, + *forma*, form, likeness.
Pepsin.	G. *pepsis*, digestion.
Pepsinogen.	G. *pepsis*, digestion, + *gennao*, I produce.
Percussion.	L. *percussio*, striking.
Pericardial.	G. *peri*, around, + *kardia*, heart.
Perichondrium.	G. *peri*, around, + *chondros*, cartilage.
Perilymph.	G. *peri*, around, + L. *lympha*, fluid.
Perimysium.	G. *peri*, around, + *mys*, muscle.
Perineum.	G. *perinaion*, perineum.
Periodontal.	G. *peri*, around, + *odous*, tooth.
Periosteum.	G. *peri*, around, + *osteon*, bone.
Peristalsis.	G. *peristaltikos*, clasping and compressing.
Peritoneum.	G. *peri*, around, + *teinein*, to stretch.
Peroneal.	G. *perone*, = L. *fibula*, pin; hence pertaining to needle-shaped leg bone.
Pes.	L. *pes*, foot.
Petrosal.	L. *petrosus*, rocky.
Petrous.	L. *petrosus*, rocky.
Phagocyte.	G. *phagein*, to eat, + *kytos*, cell.
Phalanges.	G. *phalanx*, band of soldiers; singular phalanx, originally denoted whole set of bones of a digit not just one bone.
Phallic.	G. *phallikos*, pertaining to the penis.
Phallus.	G. *phallos*, phallus; penis was a later meaning.
Pharynx.	G. *pharynx*, throat.
Phenotype.	G. *phainein*, to display, + *typos*, type.
Philtrum.	G. *philtron*, love potion, anything that awakens love.
Phonation.	G. *phone*, voice.
Photoreceptor.	G. *phos*, light, + *recipio*, I receive.
Phrenic.	G. *phren*, diaphragm.

Physic.	G. *physikos*, natural.
Physis.	G. *phyein*, to generate, hence an outgrowth.
Pia.	L. *pius*, delicate.
Pilus.	L. *pilus*, hair.
Pineal.	L. *pinea*, pine cone.
Pinna (pl. -ae).	L. *penna*, *pinna*, feather; hence, wing.
Piriform.	L. *pirum*, pear, + *forma*, shape, likeness.
Pisiform.	L. *pisum*, pea, + *forma*, shape.
Pituitary.	L. *pituita*, slime, phlegm; at one time believed to secrete a mucous material from the brain into the nose.
Placenta.	L. *placenta*, a flat cake.
Placode.	G. *plax*, anything flat.
Planta.	L. *planta*, sole of the foot.
Plantigrade.	L. *planta*, side of foot, + *gradior*, to walk.
Plasma.	G. *plasma*, something formed.
Platelet.	OF. *plate*, plate.
Platycephaly.	G. *platys*, flat, + *kephale*, head.
Platycnemia.	G. *platys*, flat, + *kneme*, knee; hence condition of side-to-side flattening of tibia giving prominence to it's anterior border.
Platymeria.	G. *platys*, flat, + *meros*, thigh; as previous for thigh.
Platysma.	G. *platysma*, plate.
Pleomorphic.	G. *pleon*, more (or many) + G. *morphos*, forms.
Pleura.	G. *pleura*, rib; i.e. related to the ribs.
Plexus (pl. -us).	L. *plexus*, plaiting, braid.
Plica.	L. *plicare*, to fold.
Pneumatic.	G. *pneumatikos*, pertaining to breath.
Pod.	G. *pous*, *podos*, foot.

Poikilocyte.	G. *poikilos*, diversified, + *kytos*, cell.
Pollex.	L. *pollex*, thumb.
Polymorphonuclear.	G. *polys*, many, + *morphe*, form,
	+ L. *nucleus*, nut; hence mixed term.
Pons.	L. *pons, pontis*, bridge.
Popliteus.	L. *poples*, ham.
Pore.	L. *porus*, passage.
Porta.	L. *porta* (pl. *-ae*), gate.
Portal.	L. *porta* (pl. *-ae*), gate.
Portio.	L. *portio*, part.
Porus.	G. *poros*, passage.
Praecordium,	L. *prae*, in front of, + *cordis*, of the heart.
Precordium.	
Pregnancy.	L. *prae*, before, + *gnascor*, to be born.
Premolar.	L. *pre*, in front, + *molaris*, molar.
Prepuce.	L. *praeputium*, foreskin.
Presbyopia.	G. *presbys*, old man, + *opsis*, sight.
Primordial.	L. *primordium*, beginning.
Procerus.	L. *procerus*, tall, extended.
Processus.	L. *processus*, going forwards.
Prochordal.	G. *pro*, in front of, + *chorde*, cord.
Proctodeum.	G. *proktos*, anus, + *hodaios*, pertaining to a way.
Profundus.	L. *profundus*, deep.
Progeria.	G. *pro*, before, + *geras*, old age.
Progesterone.	G. *pro*, before, + *gerere*, to bear.
Prognathism.	G. *pro*, in front of, + *gnathos*, jaw.
Promontory.	G. *promontorium*, mountain ridge.
Pronator.	L. *pronare*, to turn face downward.
Pronephros.	G. *pro*, before, + *nephros*, kidney.
Pronograde.	L. *pronus*, bent downward, + *gradus* step.
Prophase.	G. *pro*, before, + *phainein*, to show.
Proprioceptor.	L. *proprius*, special, + *capere*, to take.

Proptosis.	G. *pro*, before, + *ptosis*, falling.
Prosector.	G. *pro*, before, + L. *secare*, to cut.
Prosencephalon.	G. *pro*, before, + *enkephalos*, brain.
Prostate.	L. *pro*, in front, + *stare*, to stand.
Prosthion.	G. *prosthen*, before.
Proto-.	G. *protos*, first.
Protoplasm.	G. *protos*, first, + *plasma*, form.
Protuberance.	L. *protubero*, I swell.
Proximal.	L. *proximus*, next.
Psoas.	G. *psoa*, loin.
Pterygoid.	G. *pteryx*, wing, + *eidos*, likeness, shape.
Ptosis.	G. *ptosis*, fall.
Puberty.	L. *pubes*, mature.
Pubes.	L. *pubes*, mature.
Pubis (pl. **-es**).	L. *pubes*, mature.
Pudendal.	L. *pudere*, to be ashamed.
Pulmonary.	L. *pulmo*, lung.
Pulmones.	L. *pulmo*, lung.
Pulpa.	L. *pulpa*, soft part of animal body.
Pulvinar.	L. *pulvinar*, cushioned couch.
Punctum.	L. *punctum*, point.
Pupil.	L. *pupa*, girl.
Putamen.	L. *putamen*, shell, husk.
Pyelogram.	G. *pyelos*, tub, trough, + *gramma*, mark.
Pyelograph.	G. *pyelos*, tub, trough, + *graphein*, to draw.
Pyknic.	G. *pyknos*, compact.
Pyknosis.	G. *pyknos*, compact.
Pylorus.	G. *pylouros*, gate-keeper.
Pyramid.	G. *pyramis*, pyramid.
Pyriform.	L. *pirum*, pear, + *forma*, shape.

Q

Quadrangular.	L. *quattuor* four + L. *angulus*, four angles.
Quadratus.	L. *quadratus*, squared.
Quadriceps.	L. *quattuor*, four, + *caput*, head.
Quadrigeminus.	L. *quadrigeminus*, four-fold, four.
Quadruplets.	L. *quadruplaris*, fourfold.

R

Racemose.	L. *racemosus*, clustering.
Radial.	L. *radius*, rod, spoke.
Radicle.	L. *radix*, root.
Radius.	L. *radius*, rod, spoke.
Radix.	L. *radix*, root.
Ramify.	L. *ramus*, branch, + *facere*, to make.
Ramus.	L. *ramus*, branch.
Raphe.	G. *raphe*, seam.
Receptor.	L. *recipere*, to take back, receive.
Recess.	L. *recessus*, retreat.
Rectum, rectus.	L. *rectus*, straight.
Reflect.	L. *reflecto*, to turn back.
Renal.	L. *renes*, kidneys.
Restiform.	L. *restis*, rope, + *forma*, shape, form.
Rete.	L. *rete*, network.
Reticulocyte.	L. *reticulum*, little net, + *kytos*, cell.
Reticulum.	L. *reticulum*, little net.
Retina.	L. *rete*, net.
Retinaculum.	L. *retinaculum*, band.
Retraction.	L. *re*, back, + *trahere*, to draw.
Retractor.	L. *retrahere*, to draw back.
Rhinal.	G. *rhis*, nose.
Rhinencephalon.	G. *rhis*, nose, + *enkephalos*, brain.

Rhinion.	G. *rhinion*, nostril.
Rhombencephalon.	G. *rhombos*, rhomboid, + *enkephalos*, brain.
Ribosome.	G. *ribose*, an aldopentose, + *soma*, body.
Rima.	L. *rima*, cleft.
Risorius.	L. *risus*, laughter.
Rostrum (pl. -a).	L. *rostrum*, beak.
Rotator.	L. *rotare*, to whirl about.
Ruga (pl. -ae).	L. *ruga*, wrinkle.

S

Sac.	L. *saccus*, sack.
Sacculus.	L. *sacculus*, a little bag.
Sacrum.	L. *sacer*, sacred.
Sagittal.	L. *sagitta*, arrow; shape of saggital suture including the lambdoid suture.
Salivary.	L. *saliva*, saliva.
Salpinx.	G. *salpinx*, trumpet.
Saphenous.	A. *al-safin*, hidden (later G. *saphenes*, visible).
Sarcolemma.	G. *sarx*, flesh, + *lemma*, husk, skin.
Sartorius.	L. *sartor*, tailor.
Scala.	L. *scala*, staircase.
Scalene.	G. *skalenos*, uneven.
Scalp.	Teutonic. *skalp*, shell.
Scalpel.	L. s*calprum*, knife.
Scaphoid.	G. *skaphe*, boat, + *eidos*, shape, likeness.
Scapula.	L. *scapulae*, shoulder-blades.
Sciatic.	G. *ischion*, hip joint.
Sclera.	G. *skleros*, hard.
Sclerotic.	G. *skleros*, hard.
Scoliosis.	G. *skoliosis*, curvature.

Scrotum.	L. *scrotum*, skin.
Sebaceous.	L. *sebum*, tallow, grease.
Sebum.	L. *sebum*, tallow.
Sella turcica.	L. *sella*, saddle, + *turcica*, Turkish.
Semen.	L. *semen*, seed.
Semilunar.	L. *semi*, half, + *luna*, moon.
Semimembranosus.	L. *semi*, half, + *membranosus*, membrane.
Seminiferous.	L. *semen*, seed, + *ferre*, to bear.
Semitendinosus.	L. *semi*, half, + *tendinosus*, tendon.
Septum.	L. *saeptum*, fence.
Serratus.	L. *serra*, saw.
Serum.	L. *serum*, whey.
Sesamoid.	G. *sesamen*, sesame plant, or seed, + *eidos*, shape, likeness.
Sialogram.	G. *sialon*, saliva, + *gramma*, mark.
Sigmoid.	G. *sigma*, Greek letter, + *eidos*, shape, likeness.
Sinister.	L. *sinister*, left side or unlucky.
Sinus (pl. -us).	L. *sinus*, curve, cavity, bosom.
Sinusoid.	L. *sinus*, curve, cavity, + *eidos*, shape, likeness.
Situs inversus viscerum.	L. *situs*, site, position, + *inversus*, inverted, + *viscerum*, of the viscera.
Skeleton.	G. *skeletos*, dried.
Smegma.	G. *smegma*, soap.
Soleus.	L. *solea*, sandal, sole.
Soma.	G. *soma*, body.
Somatic.	G. *soma* (pl. *somata*), body.
Somatopleure.	G. *soma*, body, + *pleura*, side.
Somite.	G. *soma*, body, + suffix -*ite*, indicating origin.
Sperm.	G. *sperma*, seed.
Spermatocyte.	G. *sperma*, seed, + *kytos*, cell.
Spermatogenesis.	G. *sperma*, seed, + *genesis*, origin.

Spermatogonium.	G. *sperma*, seed, + *gone*, generation.
Spermatozoon (pl. -a).	G. *sperma*, seed, + *zoon*, animal.
Sphenoid.	G. *sphen*, wedge, + *eidos*, likeness, shape.
Sphincter.	G. *sphingein*, to bind tight.
Spina.	L. *spina*, thorn.
Splanchnic.	G. *splanchna*, viscera.
Splanchnology.	G. *splanknon*, viscus, + *logos*, treatise.
Spleen.	L. *splen*, spleen.
Spongiosum.	G. *spongia*, sponge.
Spongioblast.	G. *spongia*, sponge, + *blastos*, germ, bud.
Squamo-.	L. *squama*, scale.
Stapes.	L. *stapes*, stirrup.
Stenosis.	G. *stenosis*, narrowing.
Sternum.	G. *sternon*, chest.
Sthenic.	G. *sthenos*, strength.
Stomach.	G. *stomachos*, gullet, oesophagus.
Stomodeum, Stomatodeum.	G. *stoma*, mouth, + *odaios*, pertaining to a way.
Strabismus.	G. *strabismos*, squinting.
Stratum (pl. strata).	L. *stratum*, layer.
Stria.	L. *stria*, furrow.
Striatum.	L. *striatus*, grooved, streaked.
Stroma.	G. *stroma*, blanket.
Styloid.	G. *stylos*, pillar, + *eidos*, likeness.
Subcostal.	L. *sub*, under, + *costa*, rib.
Sublingual.	L. *sub*, under, + *lingua*, tongue.
Submandibular.	L. *sub*, under, + *mandibula*, jaw.
Substantia.	L. *substantia*, substance.
Sulcus.	L. *sulcus*, furrow.
Superciliary.	L. *super*, above, + *cilium*, eyelid.

Supination.	L. *supinus*, lying on the back.
Supinator.	L. *supinare*, to bend backward.
Supracostal.	L. *supra*, above, + *costa*, rib.
Supraspinatus.	L. *supra*, above, + *spina*, thorn.
Surgery.	L. *chirurgia*, from G. *cheir*, hand, + *ergon*, work.
Sural.	L. *sura*, calf of leg.
Sustentaculum.	L. *sustentaculum*, support.
Suture.	L. *sutura*, seam, sewing together.
Sympathetic.	G. *syn*, together, + *pathein*, to suffer.
Symphysis.	G. *syn*, together, + *physis*, growth.
Synapse.	G. *syn*, together, + *haptein*, to fasten.
Synarthrosis.	G. *syn*, together, + *arthron*, joint.
Synchrondosis.	G. *syn*, together, + *chondros*, cartilage.
Syncytium.	G. *syn*, together, + *cytos*, cell.
Syndesmosis.	G. *syn*, together, + *desmosis*, band.
Syndrome.	G. *syndrome*, occurrence.
Synergy.	G. *syn*, together, + *ergon*, work.
Synostosis.	G. *syn*, together, + *osteon*, bone.
Synovia.	G. *syn*, together, + *ovum*, egg.
Systole.	G. *systole*, contraction.

T

Tabular.	L. *tabula*, board, table.
Taenia, Tenia.	L. *taenia*, band, ribbon.
Talipes.	L. *talipedo*, weak in the feet (*talipes*, clubfoot).
Talonid.	L. *talus*, ankle bone, + G. *eidos*, form, likeness.
Talus.	L. *talus*, ankle-bone.
Tapetum.	L. *tapete*, carpet, tapestry.
Tarsus.	G. *tarsos*, sole of the foot.
Tectum.	L. *tectum* (*tego*), roof.

Tegmen.	L. *tegmen*, covering.
Tegmentum.	L. *tegumentum*, covering.
Tela.	L. *tela*, web.
Telencephalon.	G. *telos*, end, + *enkephalos*, brain.
Telophase.	G. *tele*, far, distant, + *phasis*, stage.
Temporal.	L. *temporalis*, belonging to time.
Temporal.	L. *tempora*, the temples.
Temporalis.	L. *tempus*, time.
Tendon.	L. *tendere*, to stretch.
Tensor.	L. *tendere*, to stretch.
Tentorium.	L. *tentorium*, tent.
Teres.	L. *tero*, I grind, rub.
Terminalis.	L. *terminare*, to limit.
Testicle.	L. *testiculus*, testis.
Testis.	L. *testis*, testicle; a witness.
Tetany.	G. *tetanus*, stiffness.
Thalamus.	G. *thalamos*, chamber.
Theca.	L. *theca*, envelope, sheath.
Thenar.	G. *thenar*, hand.
Theory.	G. *theoreo*, I consider, contemplate.
Thorax (pl. thoraces).	G. *thorax*, breast-plate, breast.
Thrombocyte.	G. *thrombos*, lump, + *kytos*, cell.
Thrombus.	G. *thrombus*, lump.
Thymus.	G. *thymos*, thyme.
Thyroid.	G. *thyreos*, shield, + *eidos*, form.
Tibia.	L. *tibia*, long flute.
Tonsil.	L. *tonsilla*, mooring post.
Tome.	G. *tennein*, to cut.
Trabecula (pl. -ae).	L. *trabecula*, a little beam.
Trachea.	G. *tracheia*, windpipe.
Tract.	L. *tractus*, wool drawn out for spinning.

Tragus.	G. *tragos*, goat, from small tuft of hair (goat's beard) in this region.
Trapezius.	G. *trapezion*, small four-sided table.
Triceps.	L. *tres*, three, + *caput*, head.
Tricuspid.	L. *tres*, three, + *cuspis*, point.
Trigeminus.	L. *trigeminus*, born three together.
Trigone.	L. *trigonum*, triangle.
Triquetrum.	L. *triquetrus*, three-cornered, triangular.
Trochanter.	G. *trochos*, wheel, pulley.
Trochlea.	G. *trochilia*, pulley.
Trophoblast.	G. *trophe*, nourishment, + *blastos*, bud.
Truncus.	L. *truncus*, trunk of tree.
Tuberculum.	L. *tuberculum*, a small hump.
Tuberosity.	L. *tuber*, round smooth swelling.
Tunica.	L. *tunica*, undergarment.
Tympanic.	L. *tympanum*, drum.

U

Ulna.	L. *ulna*, elbow.
Umbilicus.	L. *umbilicus*, navel.
Unciform.	L. *uncus*, hook, + *forma*, shape.
Uncinate.	L. *uncinatus*, furnished with a hook.
Uncus.	L. *uncus*, hook.
Ungual.	L. *unguis*, claw, nail.
Unipennate.	L. *unus*, one, + *penna*, feather.
Urachus.	G. *ouron*, urine, + *cheo*, I pour.
Urea.	G. *ouron*, urine.
Ureter.	G. *ouron*, urine, + *tereo*, to preserve.
Urethra.	G. *ouethra*, word invented by Hippocrates (*c.* 460 B.C.).
Urine.	G. *ouron*, urine.
Urogenital.	G. *ouron*, urine, + L. *genitalis*, genital.

Urostyle.	G. *oura*, tail, + *stylos*, pillar.
Uterus.	L. *uterus*, womb.
Utriculus.	L. *utriculus*, small skin or leather bottle.
Uvea.	L. *uva*, grape.
Uvula.	L. *uva*, grape.

V

Vagina.	L. *vagina*, sheath.
Vagus.	L. *vagus*, wandering.
Valgus.	L. *valgus*, bow-legged (**Genu valgus** now means knock-kneed).
Vallate.	L. *vallum*, rampart, walled.
Vallecula.	L. dim. of *vallus*, fossa.
Valve.	L. *valva*, leaf of door.
Valvula.	L. *valvula*, a little fold, valve.
Varus.	L. *varus*, knock-kneed (term has become transposed, see **Valgus**).
Vas.	L. *vas*, vessel.
Vascular.	L. *vasculum*, small vessel.
Vein.	L. *vena*, vein.
Velum.	L. *velum*, veil.
Venter.	L. *venter*, belly.
Ventral.	L. *venter*, belly.
Ventricle.	L. *ventriculus*, little cavity, loculus.
Vermiform.	L. *vermis*, worm, + *forma*, shape.
Vermis.	L. *vermis*, worm.
Vernix.	L. *vernatio*, shedding of snake skin.
Vertebra.	L. *vertebra*, joint.
Vertex.	L. *vertex*, whirl, whirlpool.
Vesica.	L. *vesica*, bladder.
Vesicle.	L. *vesicula*, a small bladder.
Vestibulum.	L. *vestibulum*, entrance court.

Vestige.	L. *vestigium*, trace, footprint.
Villus (pl. villi).	L. *villus*, hair.
Vincula.	L. *vinculum*, band, cord.
Visceral.	L. *viscera*, entrails, bowels.
Viscus.	L. *viscus*, internal organ.
Vitelline.	L. *vitellus*, yolk of egg.
Vitellus.	L. *vitellus*, little calf (became transferred to yolk of egg by Celsus *c.* 10 A.D.).
Vitreus, vitreous.	L. *vitreus*, of glass; hence, transparent.
Viviparous.	L. *vivus*, living, + *parere*, to beget.
Volar.	L. *vola*, palm of hand.
Volvulus.	L. *volvulus*, twisted.
Vomer.	L. *vomer*, ploughshare.
Vorticosae.	L. *vertex*, whirl, whirlpool.
Vulva.	L. *volva*, cover.

X

Xanthoderm.	G. *xanthos*, yellow, + *dermos*, skin.
Xiphoid.	G. *xiphos*, sword, + *eidos*, shape, likeness.

Z

Zona.	L. *zona*, belt, girdle.
Zonula.	L. dim. of *zona*, belt, girdle.
Zygapophysis.	G. *zygon*, yoke, + *apophysis*, process of a bone.
Zygoma.	G. *zygoma*, yoke.
Zygomatic.	G. *zygoma*, yoke.
Zygote.	G. *zygoein*, to yoke together.